This book belongs to

....

To

MY FIRST GRANDSON, YEONHAEJOO

FOR WHOSE USE IT WAS ORIGINALLY DESIGNED,

THIS

"MATH AND MINDFULNESS BOOK"

IS DEDICATED

BY

HIS AFFECTIONATE GRANDMOTHER,

THE AUTHOR

contents

Try out the activities in any order.

Activities get more challenging towards the end of the text.

ACTIVITY	PAGES

Thank you for purchasing this
Math and Mindfulness book.

We hope this book brings you hours of fun!

This book includes free
bonus content and math games
which are available at
acorns2oakes.com

EVERY DAY MEANS SOMETHING

Develop a Mindful Morning Mantra

I will do things at my own pace

I will not give up!

I am the author of my own life!

I got goals!

I can be different

I can ask for help

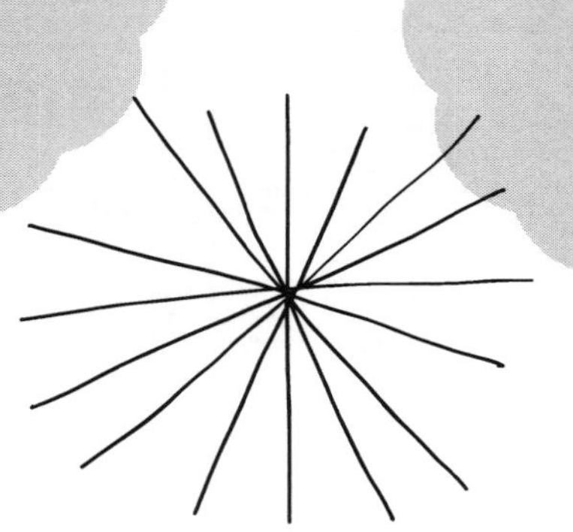

Activity 1
K-4 SOCIAL EMOTIONAL LEARNING

I can demonstrate an awareness of my personal strengths, challenges and aspirations.

ABOUT YOU

Draw your self portrait

My favourite thing to do is...

I hope that....

Someone who makes me happy is ...

My biggest success has been....

Activity 2
K-4 COUNTING & CARDINAILTY

I can understand and connect counting to the number of items in a set.
I can rewrite addition as multiplication.
I can work carefully and check my work.

ICE CREAM CONES

DRAW 2 SCOOPS OF ICE CREAM

DRAW 3 SCOOPS OF ICE CREAM

DRAW 4 SCOOPS OF ICE CREAM

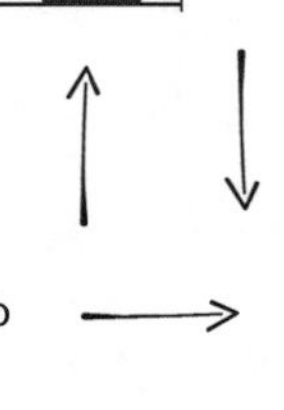

Can you find at least three different shortest distance paths the race car can take to get to the trophy? Add one directional arrow to each box to show the shortest path to follow to earn the trophy. Use a different color for each path.

↑ ↓ → ←

REWRITING ADDITION AS MULTIPLICATION

Find the answer and rewrite each addition problem as a multiplication problem.

ADDITION PROBLEM	ANSWER	MULTIPLICATION PROBLEM
2 + 2 + 2 + 2 + 2 + 2	12	2 X 6
3 + 3 + 3 + 3 + 3		
4 + 4 + 4 + 4 + 4 + 4 + 4		

Activity 3
K-3 GEOMETRY

I can reason with shapes and their attributes.
I can extend a sequence,
I can explain my thinking and try to understand the thinking of others.

Activity 4-5
1-4 OPERATIONS AND ALGEBRAIC THINKING

I can count within a 100.
I can interpret products of whole numbers in situations involving equal groups or arrays.

WHAT DO YOU SEE?

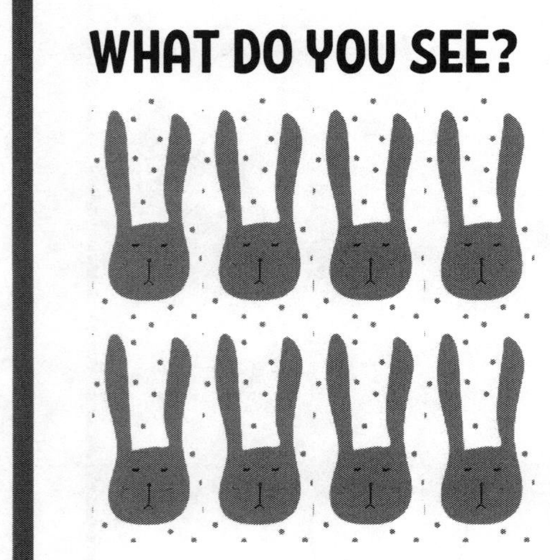

In the space above, write what you see in this box.

Activity 6
2-4 OPERATIONS AND ALGEBRAIC THINKING

I can interpret products of whole numbers in situations involving equal groups or arrays.
I can use a small grid to help enlarge an image to a large grid.

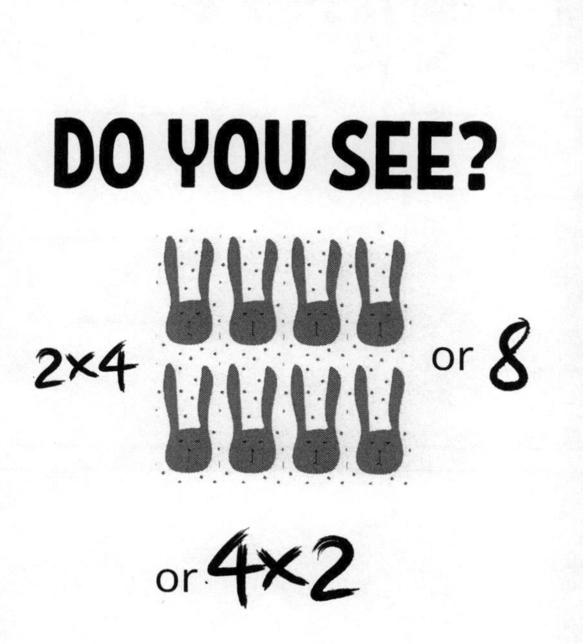

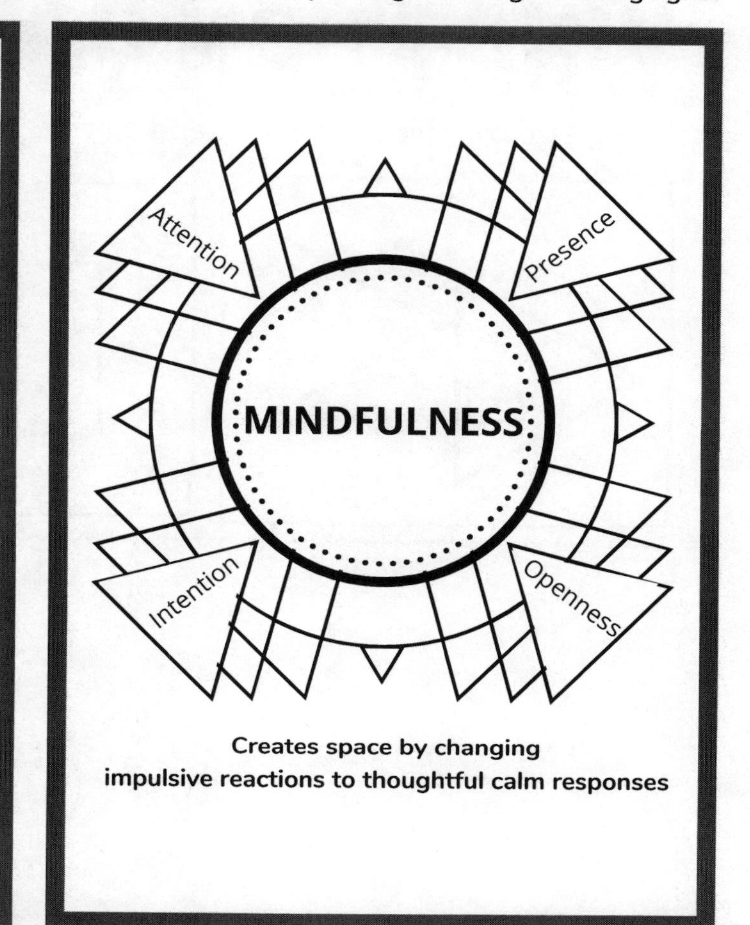

Creates space by changing
impulsive reactions to thoughtful calm responses

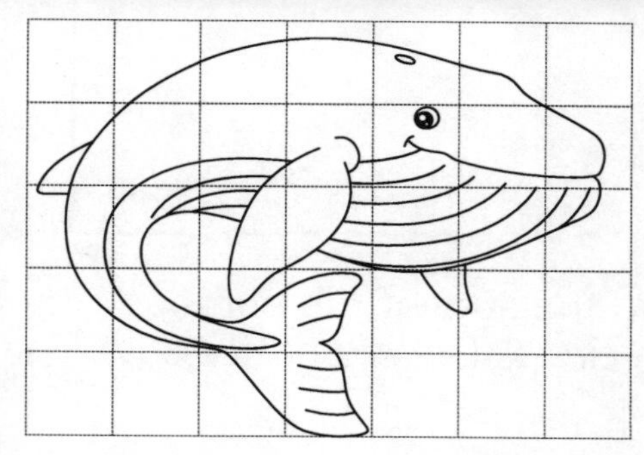

Use the two grids to help you draw a bigger whale. It may be easier to draw the whale by counting and drawing one square at a time.

Activity 7-8
1-4 CREATIVE PROBLEM SOLVING

I can hold a growth mindset and solve open ended problems.
I can look for patterns and use critical reasoning skills.
I can solve problems without giving up.

WHAT CAN YOU CREATE?

Using two triangles, a square, and a rectangle, what can you create?

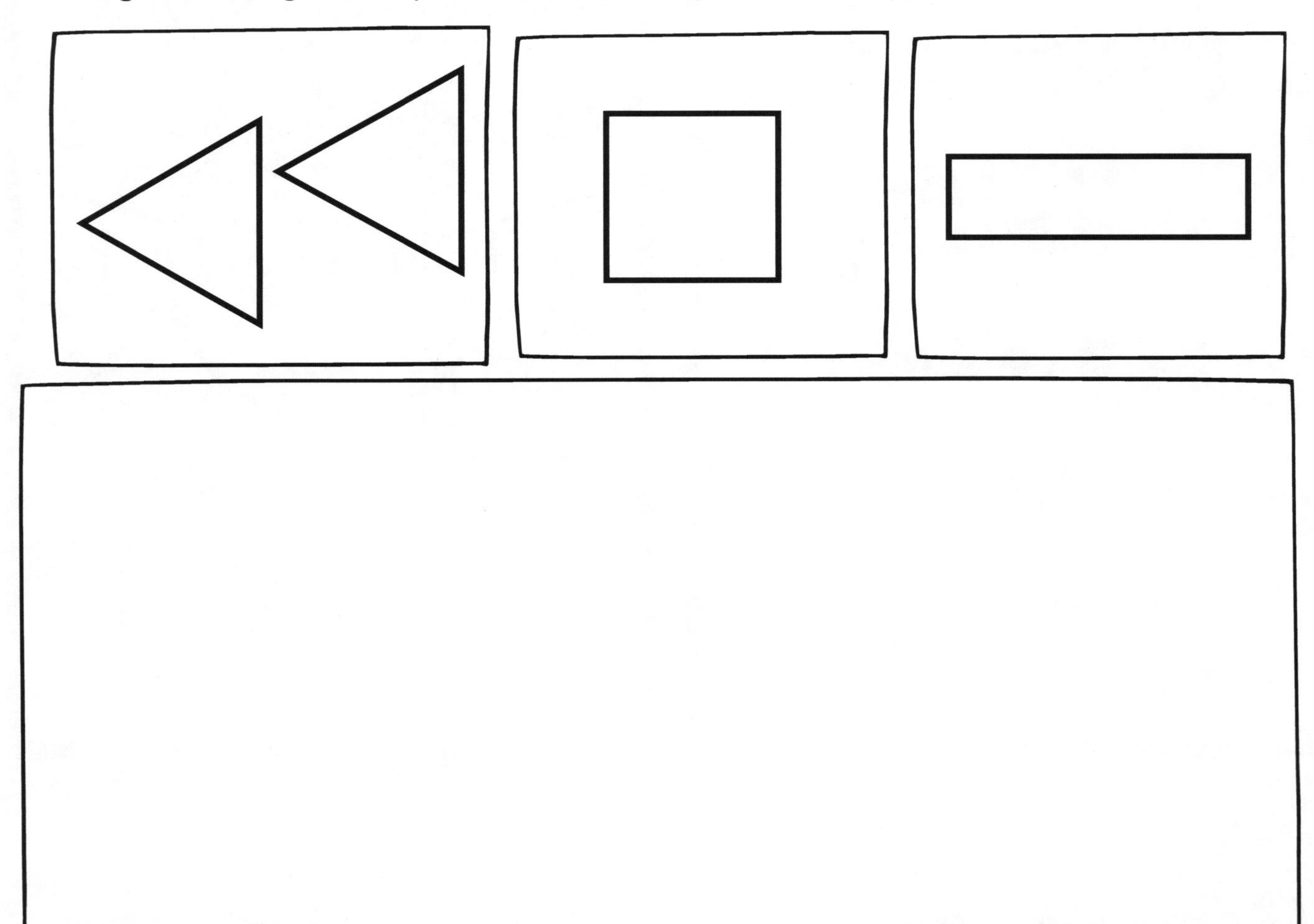

OLYMPIC RINGS

Can you find the sum of the numbers within each circle? Do you notice a pattern?

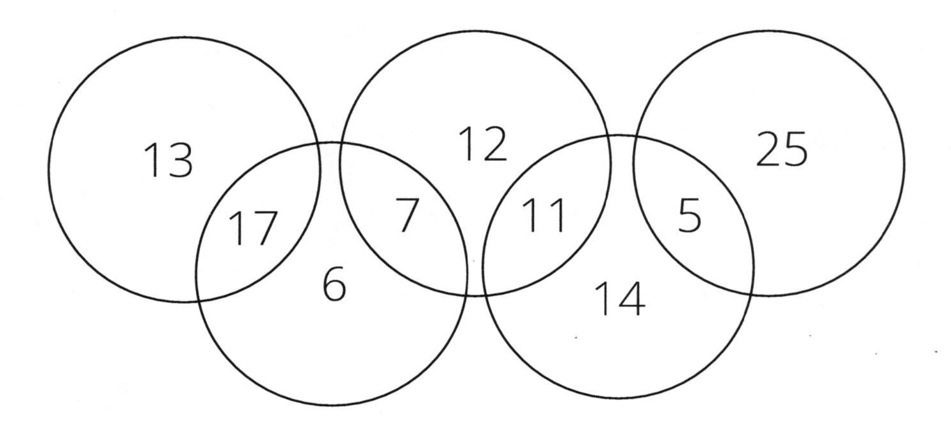

Activity 9
1-4 CREATIVE PROBLEM SOLVING

I can hold a growth mindset and solve open ended problems.
I can look for patterns and use critical reasoning skills.
I can solve problems without giving up.

MORE OLYMPIC RINGS

The sum of the numbers within each circle of the olympic ring sets is given. Use all the digits 1 to 9 to replace the letters locations A to I in the circles below with numbers. Create circles that each total to 11, 13 and 14, using only the digits 1 to 9 once for each set. Circles with sums of 12 and 15 are impossible.

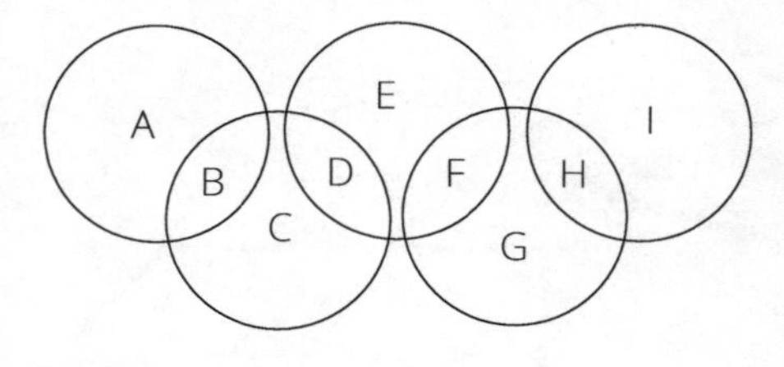

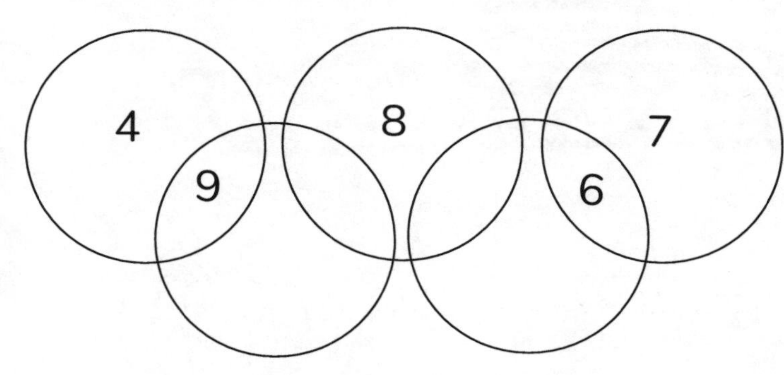

Circles each with a sum of 13

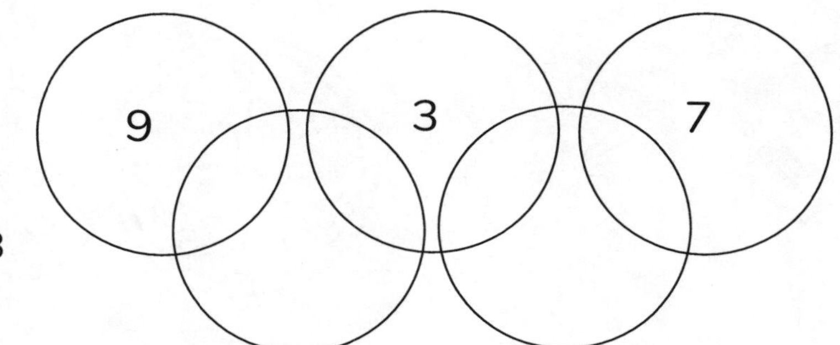

Circles each with a sum of 13

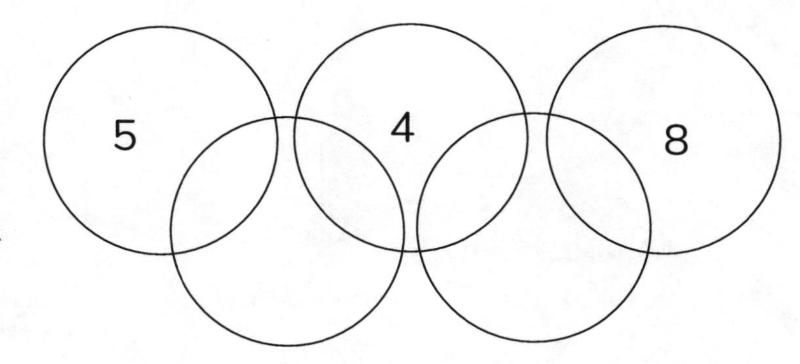

Circles each with a sum of 14

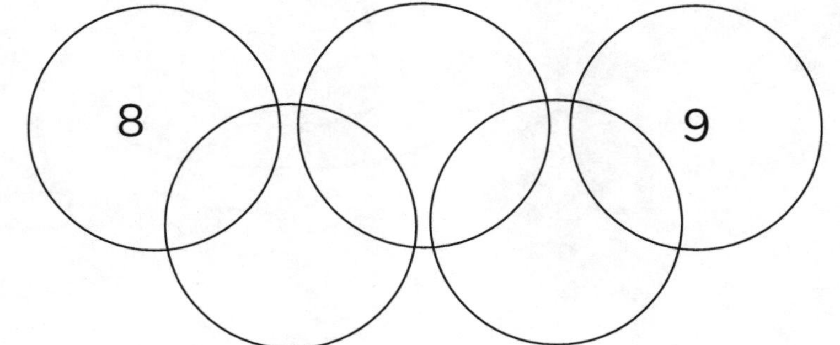

Circles each with a sum of 11

Activity 10
1-4 OPERATIONS & ALGEBRAIC THINKING

I can name all 26 images in order alphabetically.
I can think about numbers in many ways.

SEQUENCES

Can you name the 26 things in order from A to Z?

Activity 10
1-4 OPERATIONS & ALGEBRAIC THINKING

I can name all 26 images in order alphabetically.
I can think about numbers in many ways.

SEQUENCES

Write the names of the 26 images in order from A to Z?

1. A
2. B
3. C
4. D
5. E
6. F
7. G
8. H
9. I
10. J
11. K
12. L
13. M
14. N
15. O
16. P
17. Q
18. R
19. S
20. T
21. U
22. V
23. W
24. X
25. Y
26. Z

Activity 11
1-4 PROBLEM SOLVING

I can systematically solve a problem by finding the route out of the maze.
I can use what I know to solve new problems.

Choose your own path

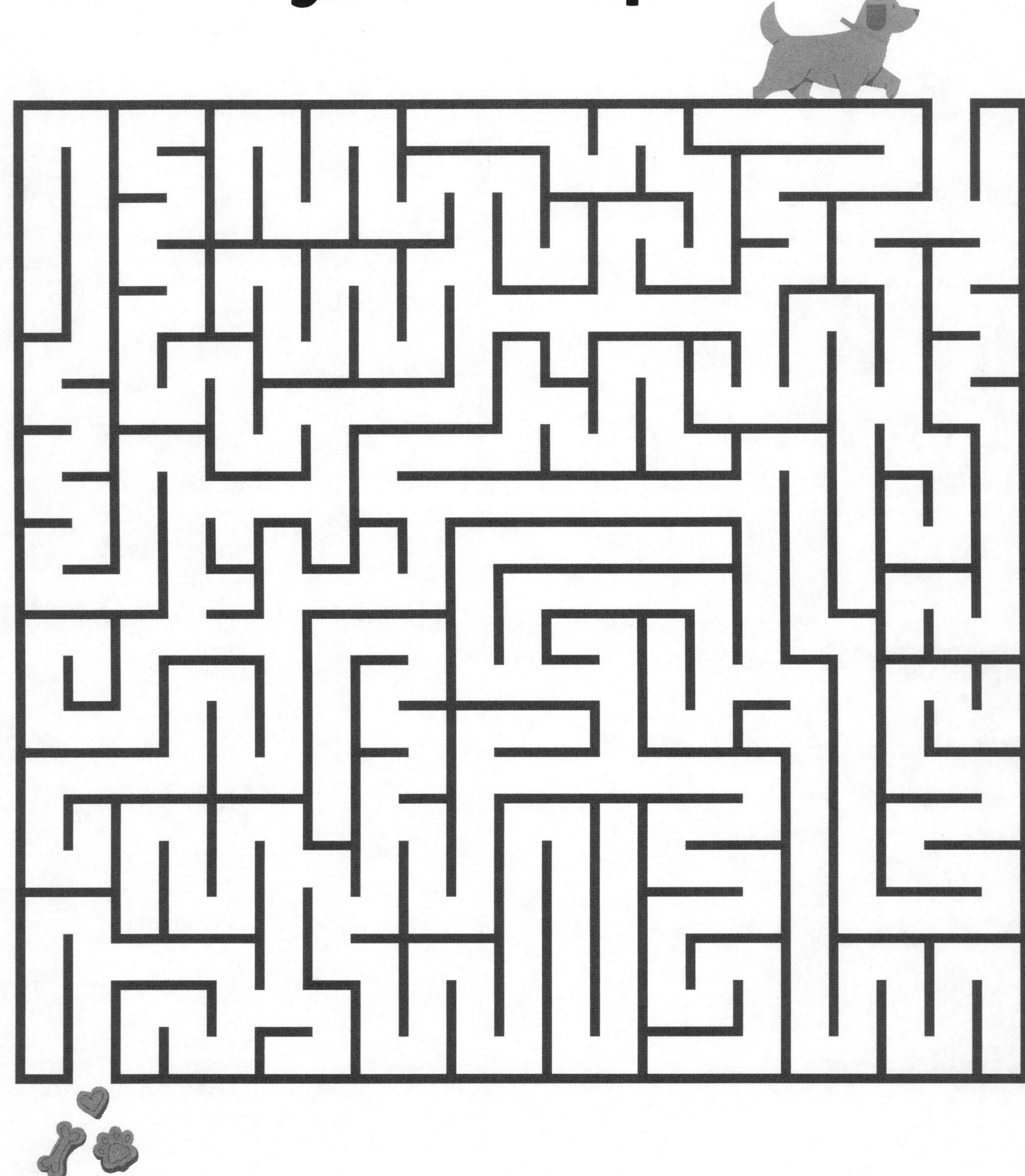

Activity 12
K-2 OPERATIONS AND ALGEBRAIC THINKING

I can add and subtract within 20.
I can think about numbers in many ways.
I can solve problems without giving up.

MAGIC TRIANGLES

Take a pair of scissor, cut out each circle, use these numbers as tools to help you solve the magic triangles.

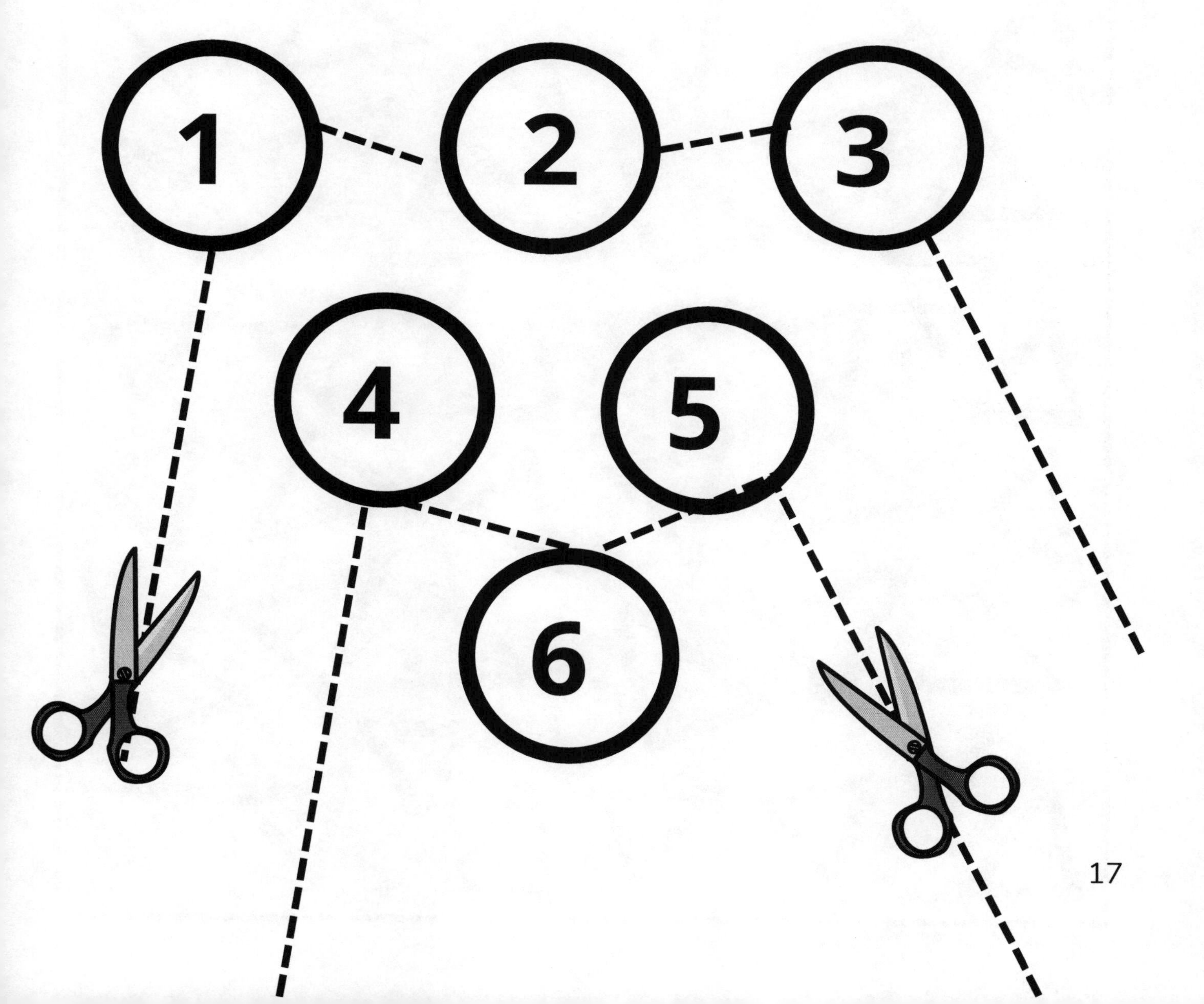

Activity 12
K-2 OPERATIONS AND ALGEBRAIC THINKING

I can add and subtract within 20.
I can think about numbers in many ways.
I can solve problems without giving up.

MAGIC TRIANGLES

Place each number 1 through 6 in the circles once around the outside of a triangle, making the numbers on each side of the triangle add up to the middle number.

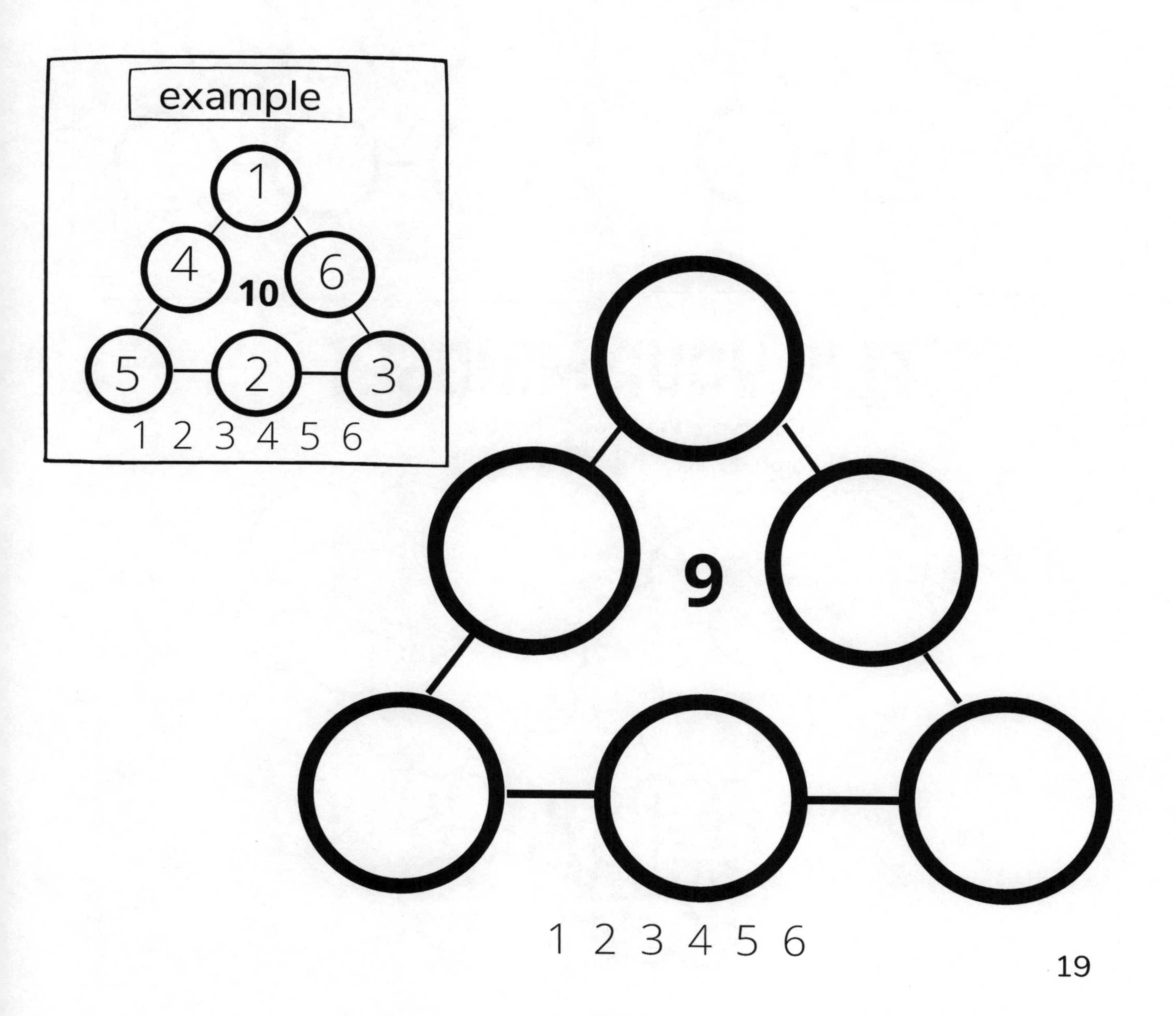

Activity 12-13
K-2 OPERATIONS AND ALGEBRAIC THINKING

I can add and subtract within 20.
I can think about numbers in many ways.
I can solve problems without giving up.

MAGIC TRIANGLES

Place each number 1 through 6 in the circles once around the outside of a triangle, making the numbers on each side of the triangle add up to the middle number.

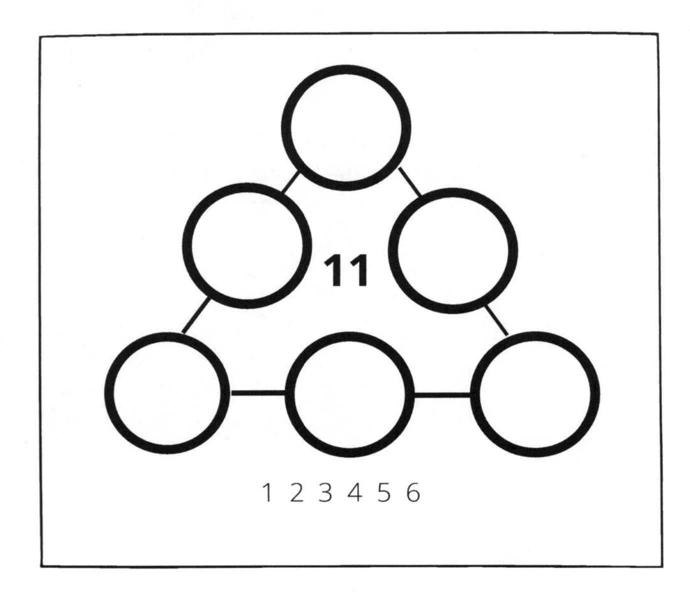

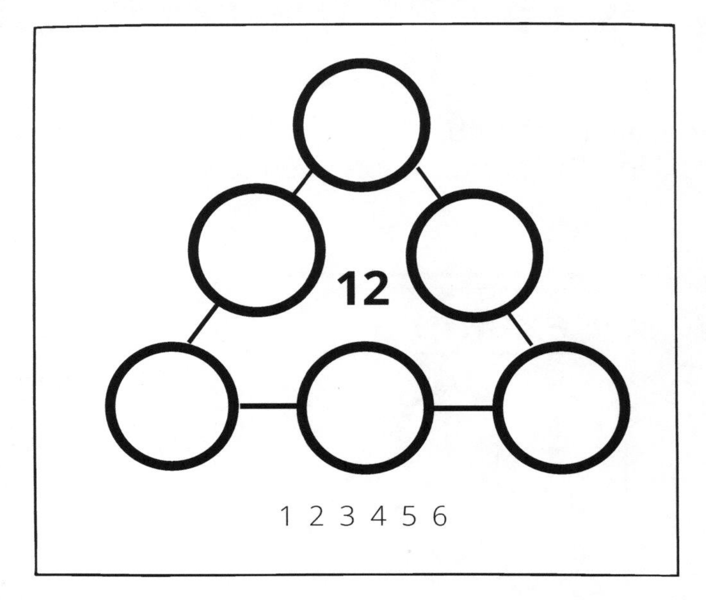

FLIP YOUR MINDSET

Complete the Speech Bubbles

Instead of:

Instead Say:

MORE MAGIC TRIANGLES

Take a pair of scissor, cut out each circle, use these numbers as tools to help you solve more magic triangles.

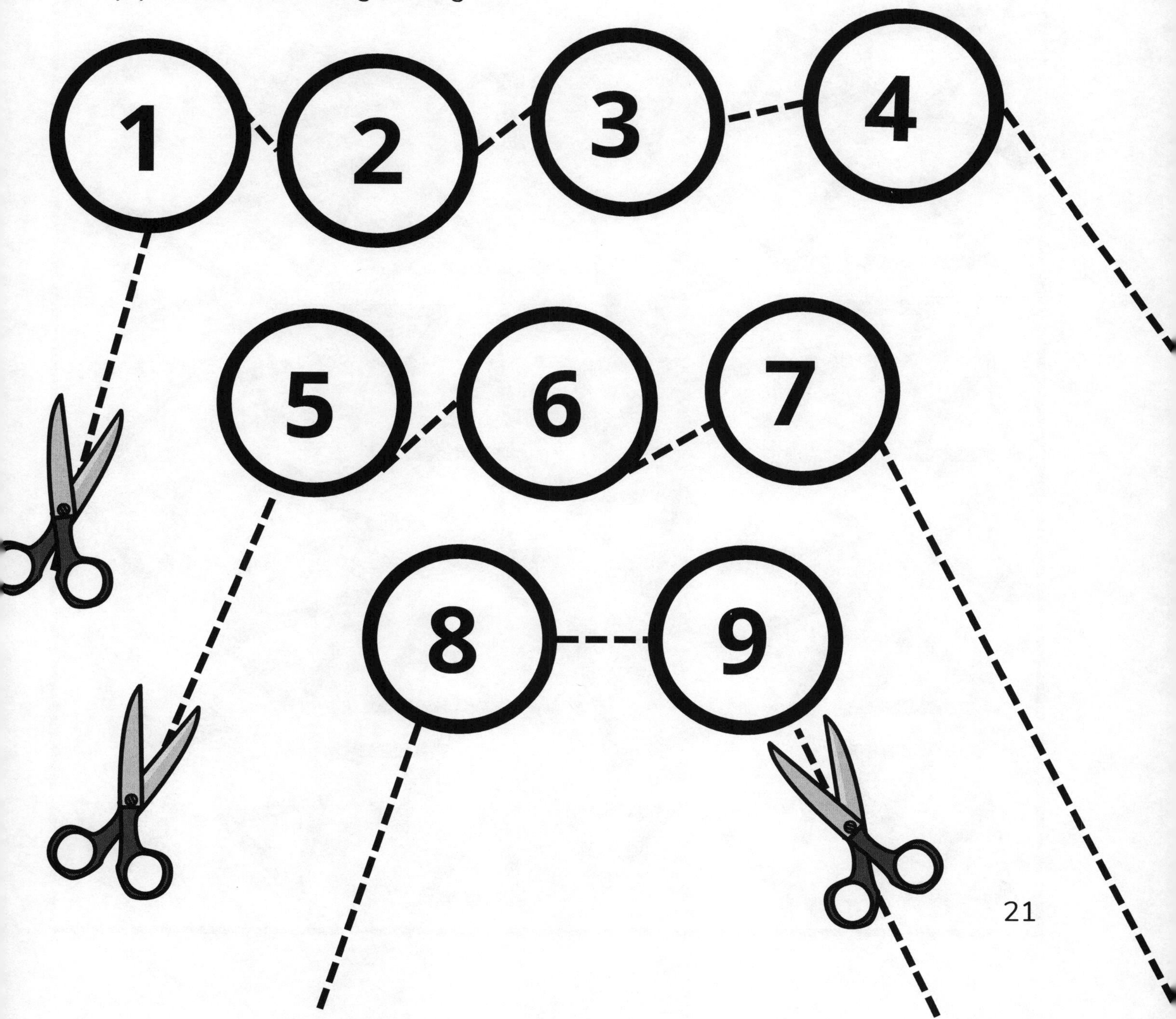

I can fluently add within 20 using mental strategies.
I can solve problems without giving up.

MORE MAGIC TRIANGLES

Try placing each number 1 through 9 around the outside of a triangle, making the numbers on each side of the triangle add up to the middle number.

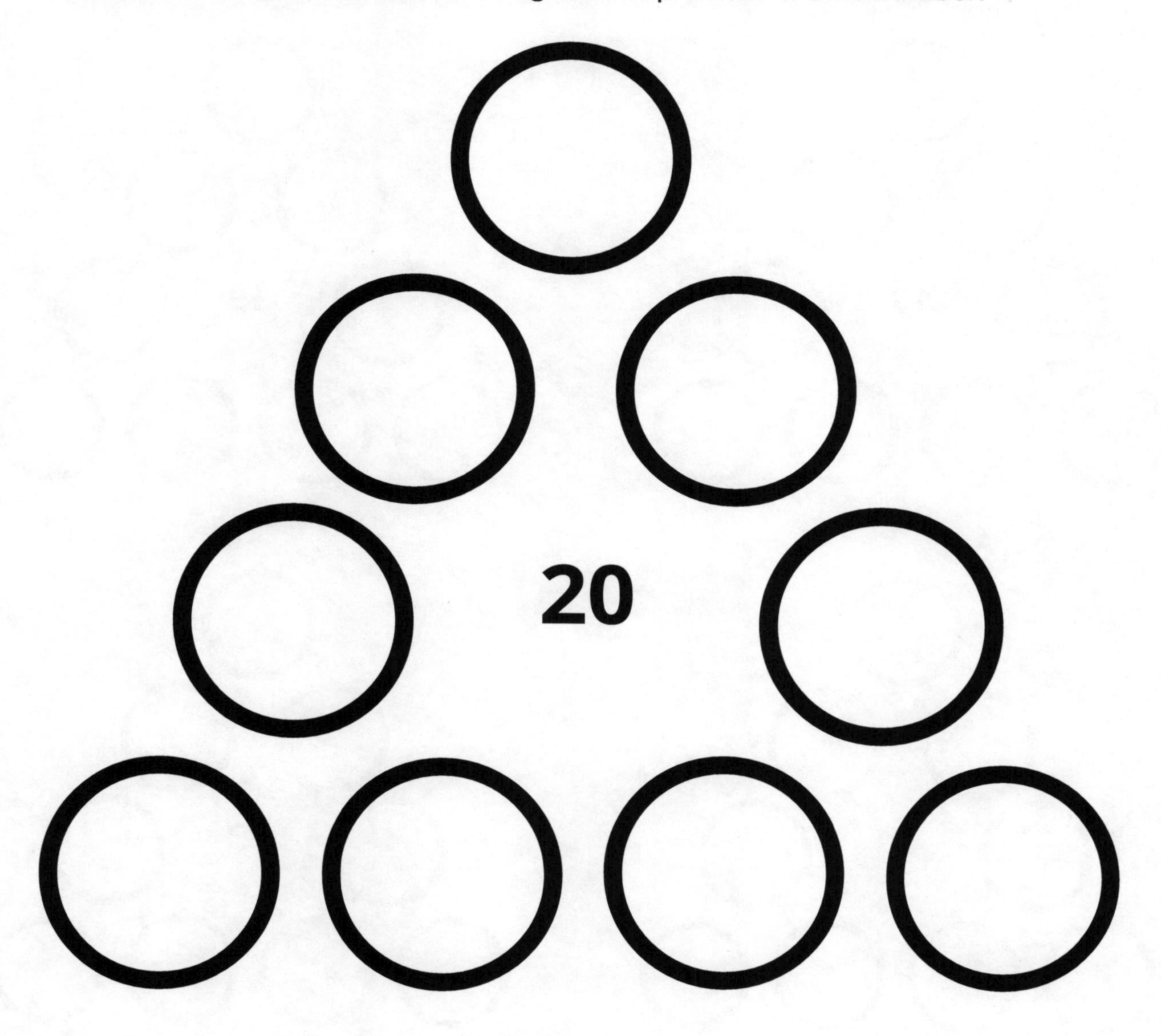

1 2 3 4 5 6 7 8 9

I can fluently add within 20 using mental strategies.
I can solve problems without giving up.

MORE MAGIC TRIANGLES

Try placing each number 1 through 9 once around the outside of the triangle, making the numbers on each side of the triangle add up to the middle number.

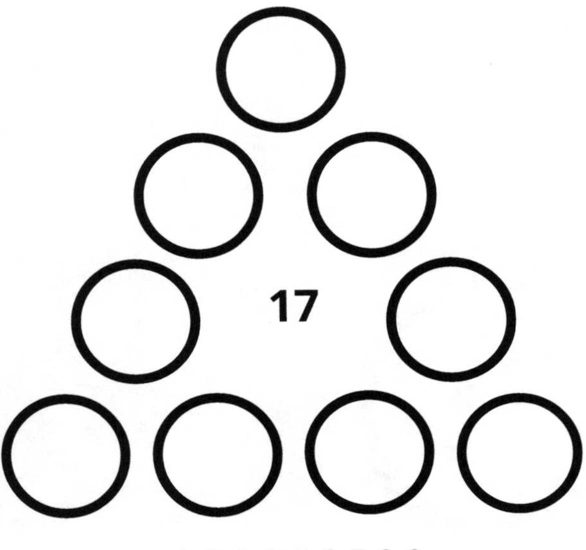

1 2 3 4 5 6 7 8 9

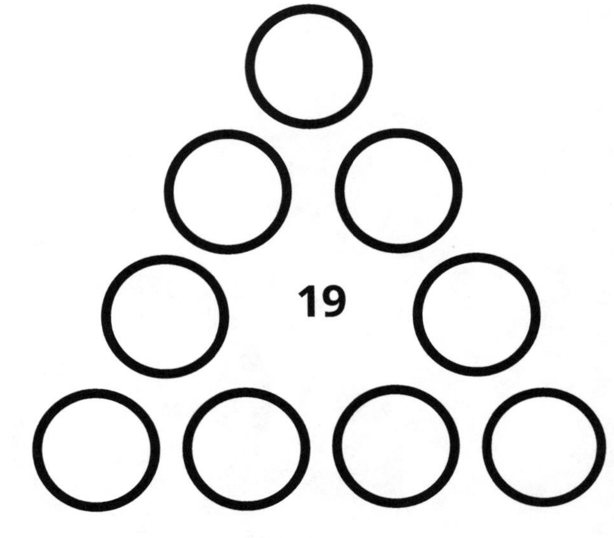

1 2 3 4 5 6 7 8 9

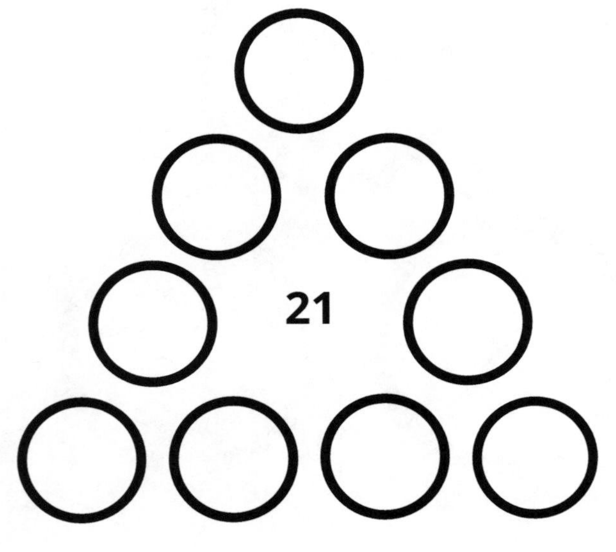

1 2 3 4 5 6 7 8 9

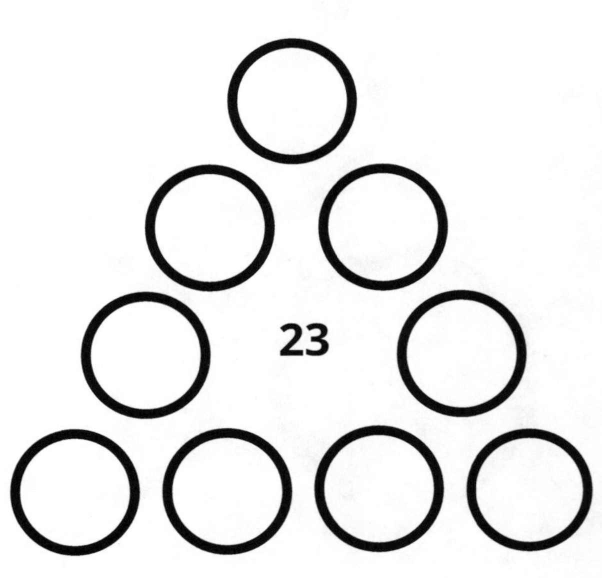

1 2 3 4 5 6 7 8 9

I can focus on how to choose and apply color in a design to bring an awareness to the present moment.

If something happens that upsets you.

Be cool & Let it go!

Activity 16
1-3 OPERATIONS AND ALGEBRAIC THINKING

I can skip count by 2s.
I can work carefully and check my work.

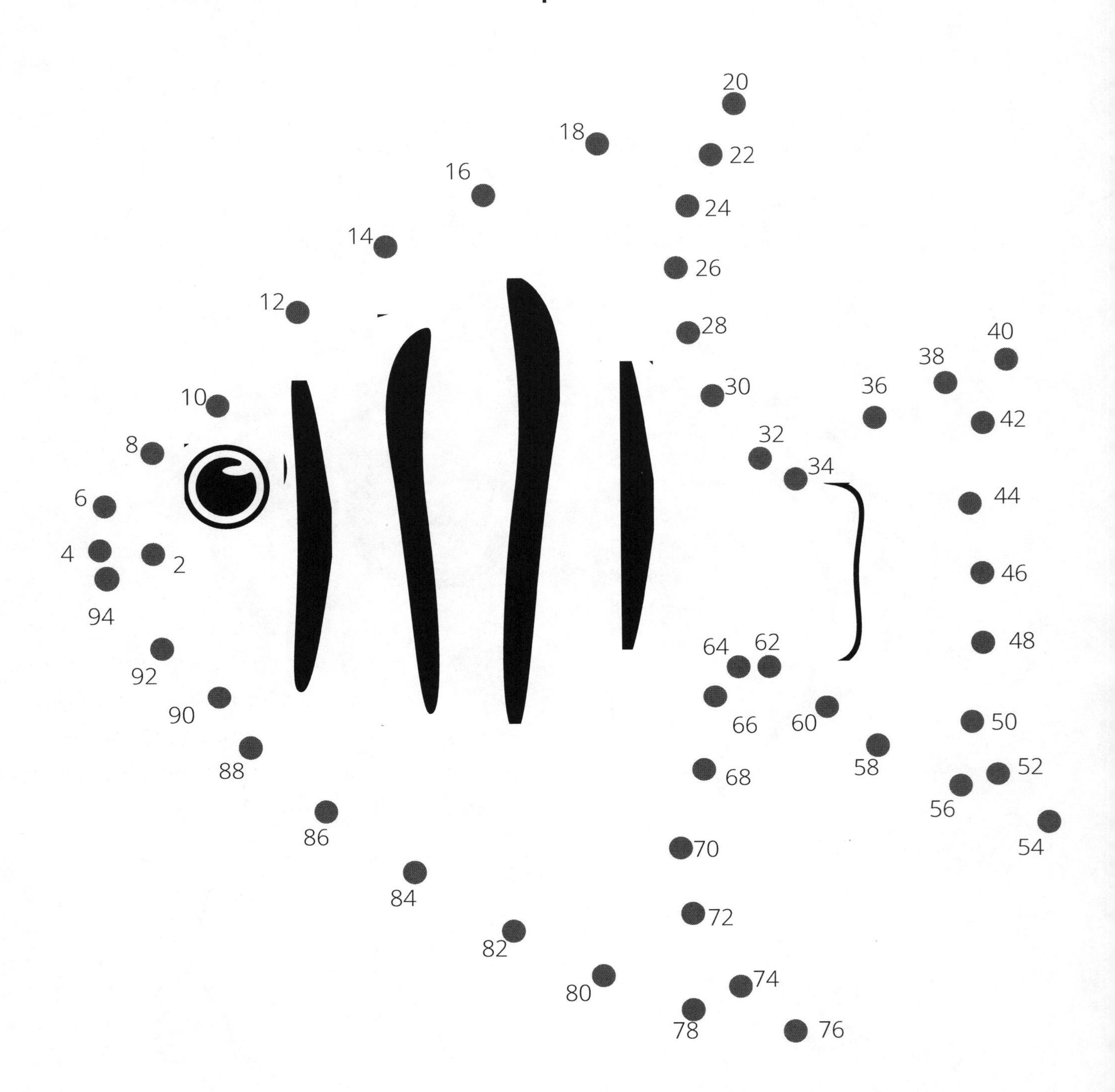

Take care of your body
Drink More Water!

Activity 17-18
1-4 SOCIAL EMOTIONAL LEARNING
OPERATIONS AND ALGEBRAIC THINKING

I can fluently multiply a one digit by a one digit.
I can demonstrate an awareness of my own emotions and external supports.

Color the array

Color the number of rows by the number of columns, to solve the following multiplication problem **3 x 8 =**

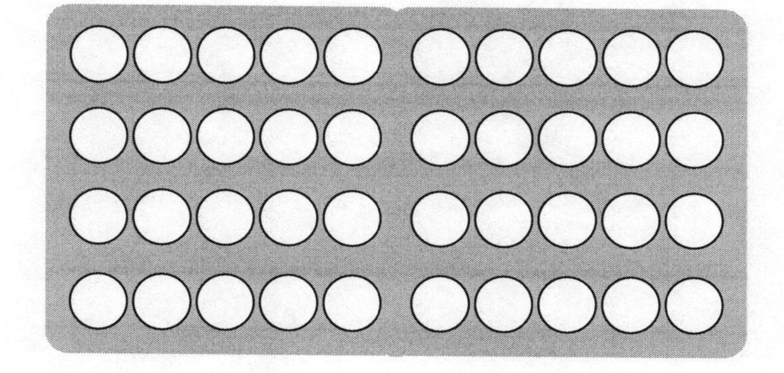

GRATITUDE FLAG

Complete the thought with positive affirmations with a drawing or words. Cut out the flag and tape to the edge of a pencil. Spin the flag, by placing the the pencil between the palms of your hands, then rubbing the palms of your hands together forward and backward.

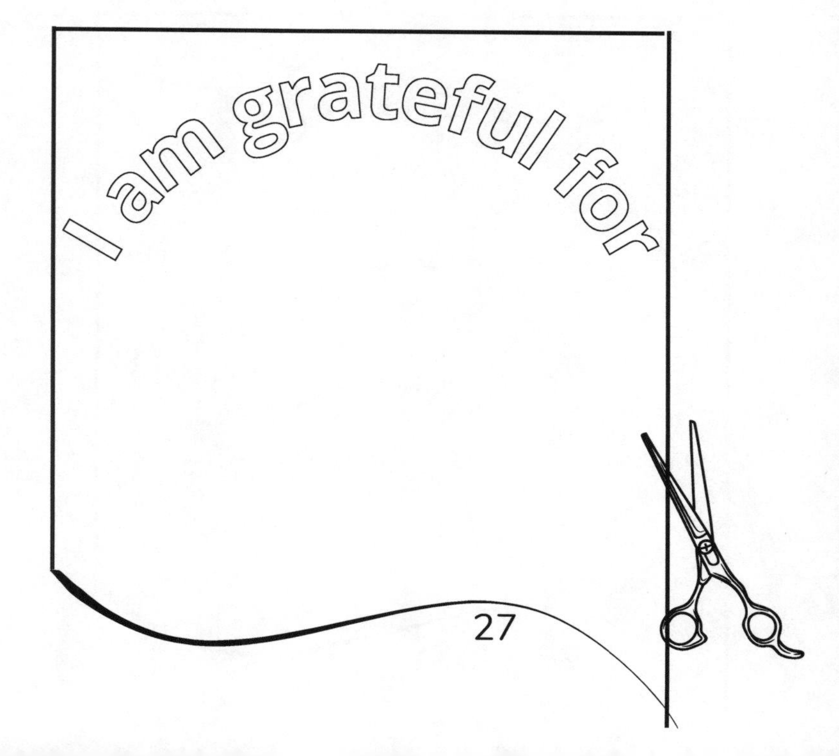

Activity 18-19
K-4 SOCIAL EMOTIONAL LEARNING
GEOMETRY

I can identify the parts of a whole and recreate a recognizable image.
I can demonstrate an awareness of my own emotions and external supports.

GRATITUDE FLAG

Complete the thought with positive affirmations with a drawing or words. Cut out the flag and tape to the edge of a pencil. Spin the flag, by placing the the pencil between the palms of your hands, then rubbing the palms of your hands together forward and backward.

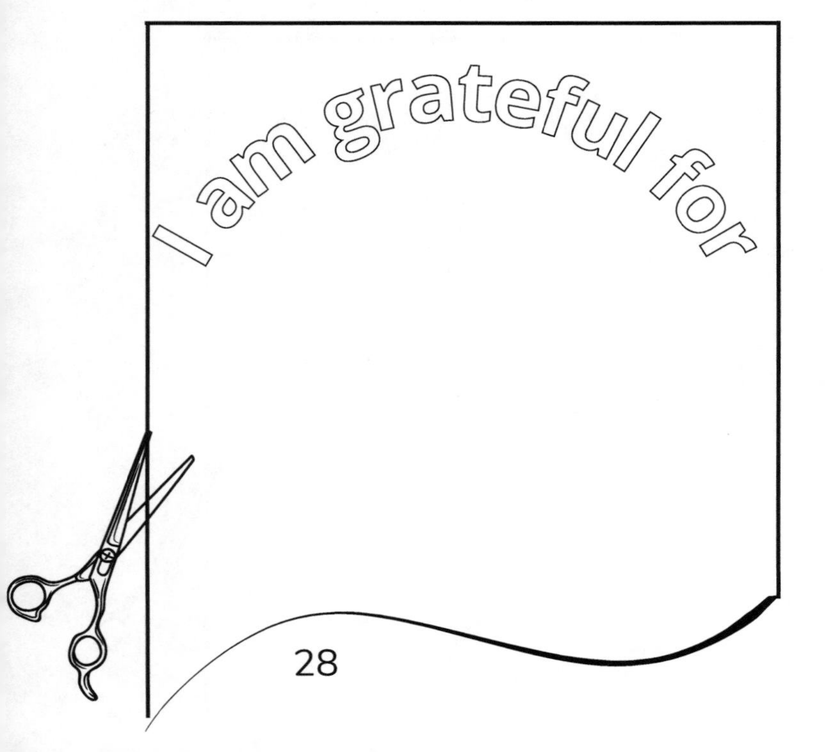

Copy the figure below onto the larger grid.

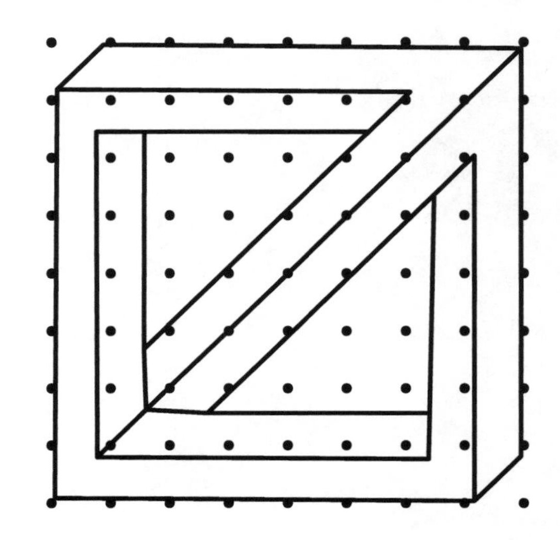

Activity 20
3-4 OPERATIONS AND ALGEBRAIC THINKING

I can find the products of a one digit times a one digit.
I can think about numbers in many ways.

MULTIPLICATION MISTAKES

There are 24 mistakes in the multiplication chart! How many can you find? Color in the circles with the errors. Write the correct equations on the left.

X	1	2	3	4	5	6	7	8	9	10
1	1	2	4	4	5	6	7	8	9	10
2	2	4	6	8	10	11	14	16	18	20
3	3	6	9	12	16	18	21	25	27	33
4	4	8	12	16	21	24	29	33	34	40
5	5	10	15	18	25	29	35	36	43	50
6	6	21	18	22	30	36	40	48	52	60
7	7	14	20	28	37	42	49	57	63	70
8	8	16	23	32	40	48	56	65	72	80
9	9	19	27	37	45	54	63	72	81	90
10	10	20	30	40	55	60	70	80	90	100

1
2
3
4
5
6
7
8
9
10
11
12
13
14
15
16
17
18
19
20
21
22
23
24

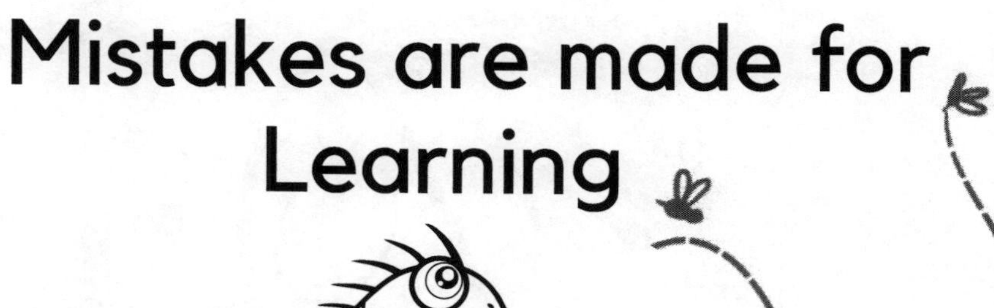

Activity 21
3-4 OPERATIONS AND ALGEBRAIC THINKING

I can count skip count by 5s.
I can work carefully and check my work.

YOU ARE BEAUTIFUL
YOU ARE LOVED

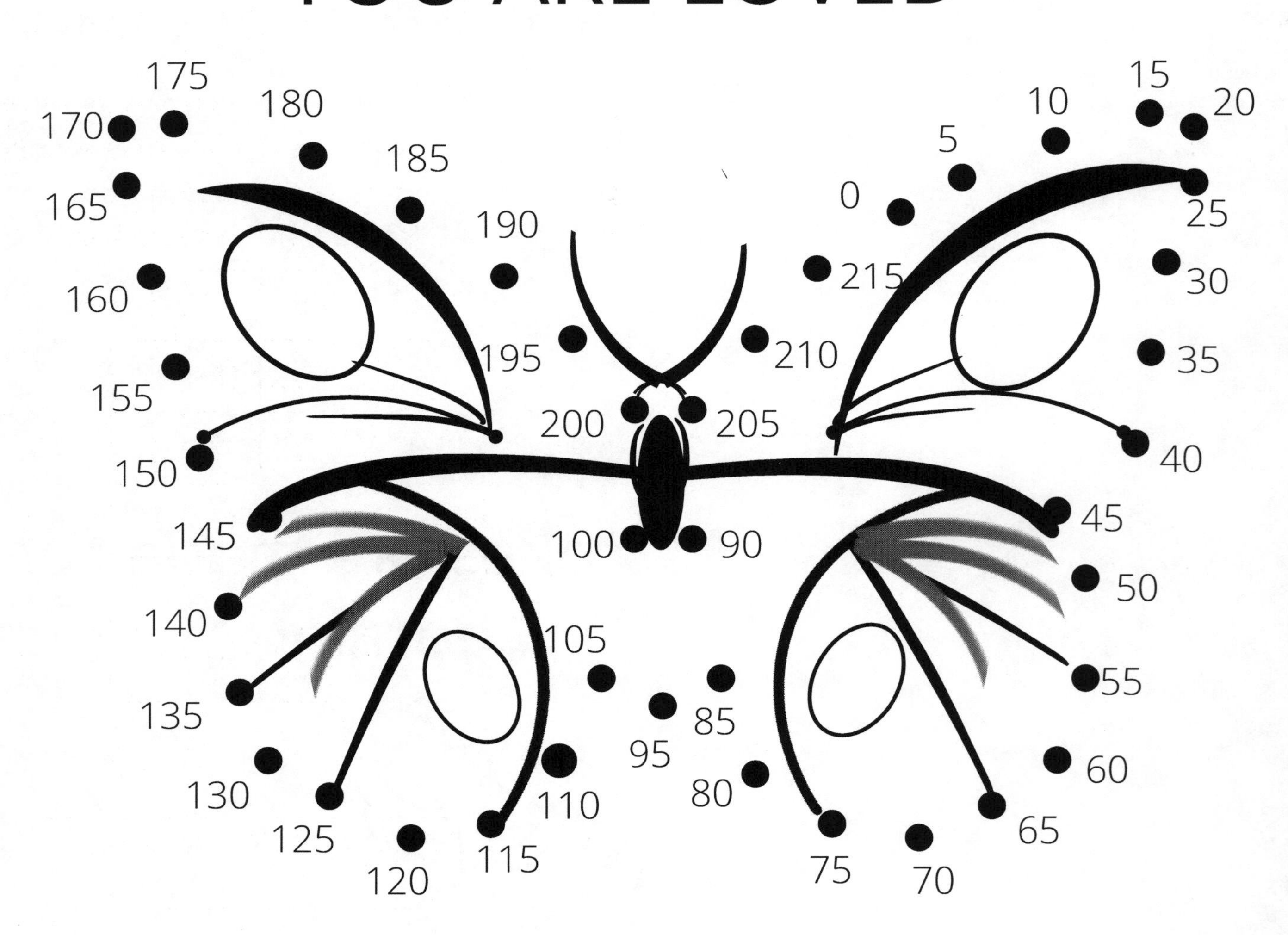

Activity 22-24
1-4 OPERATIONS AND ALGEBRAIC THINKING

I can solve problems by looking at rules and patterns.
I can fluently multiply two one digit numbers.
I can solve problems without without giving up.

CRACK THE CODE

Cryptography is a term used in computer science to keep information secret. A substitution cipher has been used here where the letters of the alphabet have been replaced with numbers. Use the key to decipher the coded message below. Then use this key to create your own coded message to share with a friend or someone else.

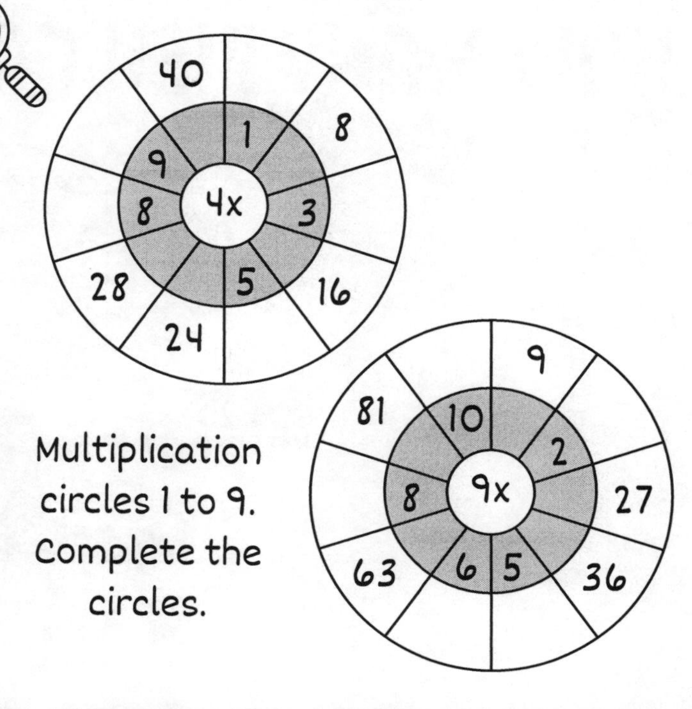

Multiplication circles 1 to 9. Complete the circles.

A	B	C	D	E	F	G	H	I	J	K	L	M
26	1	2	3	4	5	6	7	8	9	10	11	12

N	O	P	Q	R	S	T	U	V	W	X	Y	Z
13	14	15	16	17	18	19	20	21	22	23	24	25

___ ___ ' ___   ___ ___ ___ ___
8 19 18   14 10 26 24

___ ___   ___ ___ ___
19 14   26 18 10

___ ___ ___   ___ ___ ___ ___
5 14 17   7 4 11 15

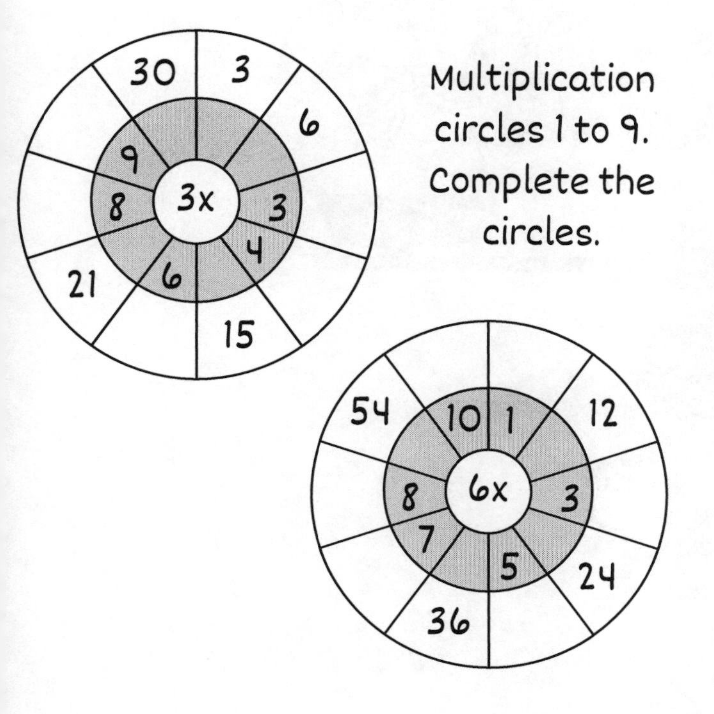

Multiplication circles 1 to 9. Complete the circles.

DECODE THE MESSAGE

A polybius square is another substitution cipher. Can you decode the secret message?

🔓	1	2	3	4	5
1	A	B	C/K	D	E
2	F	G	H	I	J
3	L	M	N	O	P
4	Q	R	S	T	U
5	V	W	X	Y	Z

W = 52   A = 11   U = 45

22-34---44-34---11---35-11-42-13---34-42---

21-34-42-15-43-44---44-34-----

21-15-15-31----42-15-43-44-34-42-15-14---

I can interpret quotients of whole numbers.
I can use what I know to solve new problems.

HOW MANY FISH CAN ONE BEAR EAT?

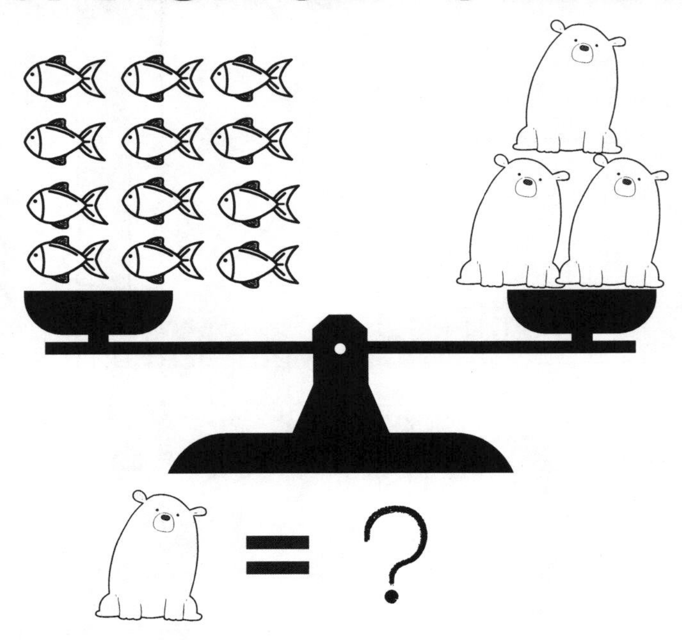

= ?

IF THE TWO SQUIRRELS CAN SHARE EQUALLY, HOW MANY ACORNS WILL EACH SQUIRREL HIDE?

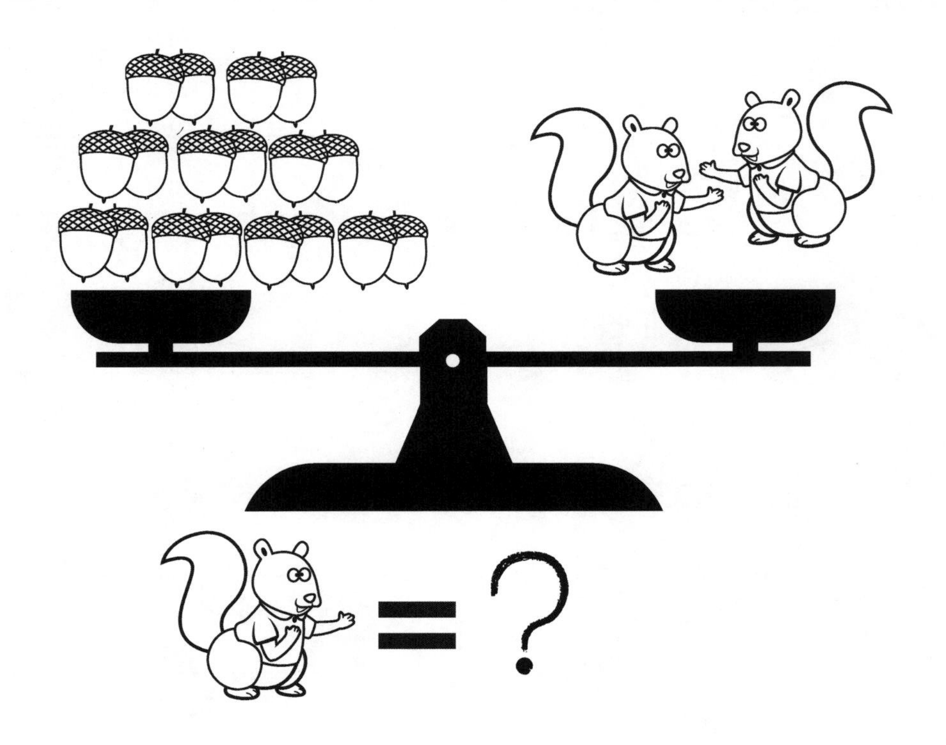

I can find sums, differences, products and quotients of whole numbers.
I can think about numbers many ways.

UNDER the SEA SUM PUZZLE

The sum of the animals in each row are given. Find and write the numerical value of each picture and the the question mark.

			= 30
			= 18
			= 2
			= ?

= ____ = ____ = ____ ?= ____

ANIMAL SUM PUZZLE

The sum of the animals in each row and column in the square are given. Find and write the value of each animal, in the spot provided. The bunny has a value less than eight.

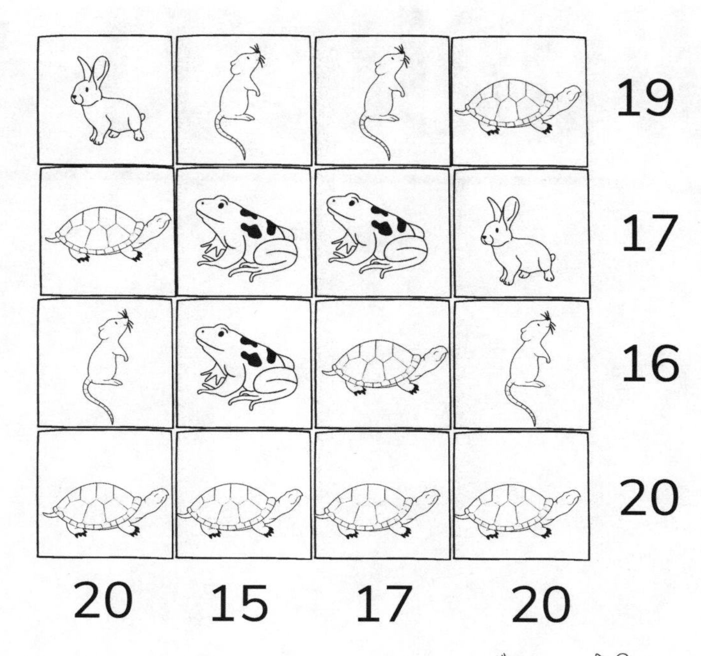

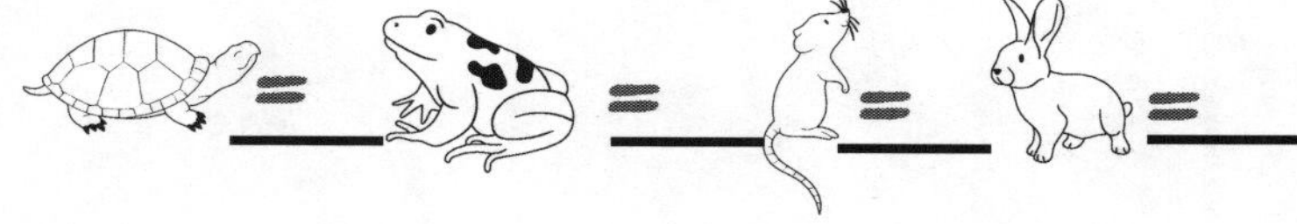

"Take in a deep breath,
inhale count 1-2-3-4,then
exhale count 1-2-3-4
and try all over
again."

Activity 27
1-4 CREATIVE PROBLEM SOLVING

I can hold a growth mindset and solve open ended problems.
I can solve problems without giving up.

IMAGINATION WORKOUT #1

Imagine what the design could be, then finish the drawing yourself!
Give your drawing a title.

Activity 28
1-4 PROBLEM SOLVING

I can systematically solve a problem by finding the route out of the maze.
I can use what I know to solve new problems.

BELIEVE IN YOU

EVERY
THOUGHT YOU THINK
COUNTS

EVERY
POSITIVE THOUGHT
BRINGS GOOD INTO
YOUR LIFE

EVERY
NEGATIVE THOUGHT
PUSHES GOOD AWAY

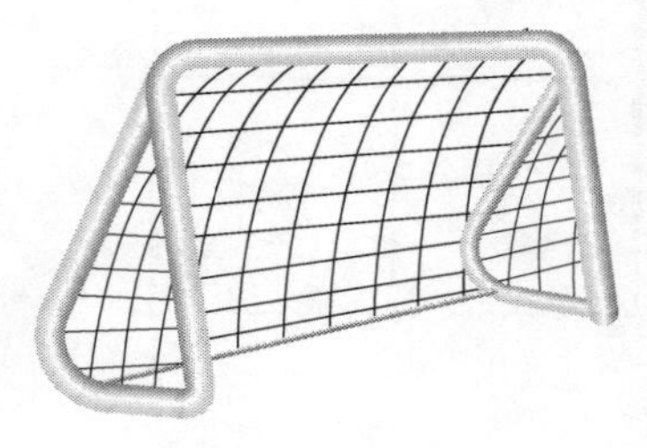

Activity 29
1-4 SOCIAL EMOTIONAL LEARNING

I can demonstrate an awareness and understanding of various emotions,

EVERYONE HAS FEELINGS

Match and write the correct emotion below each monster.

happy

proud

sad

afraid

mad

confused

Match and write the correct emotion below each monster.

annoyed

nervous

worried

angry

scared

joyful

Activity 30
1-4 PROBLEM-SOLVING

I can solve problems without giving up.

PAPERFOLDS

1

VERSION 1

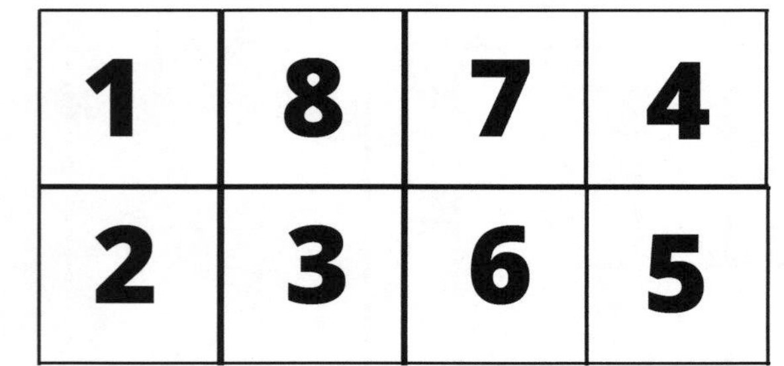

1	8	7	4
2	3	6	5

VERSION 2

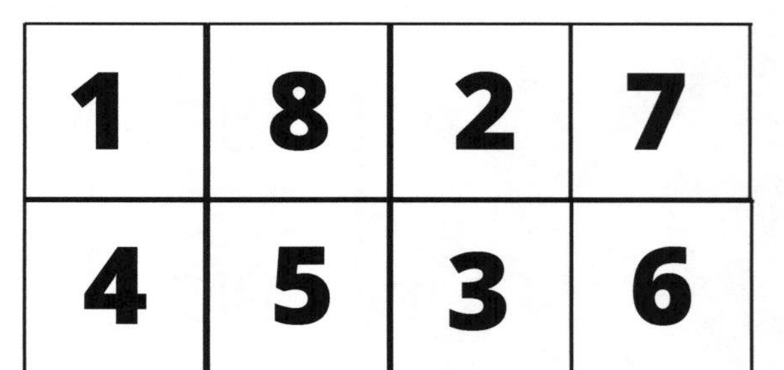

1	8	2	7
4	5	3	6

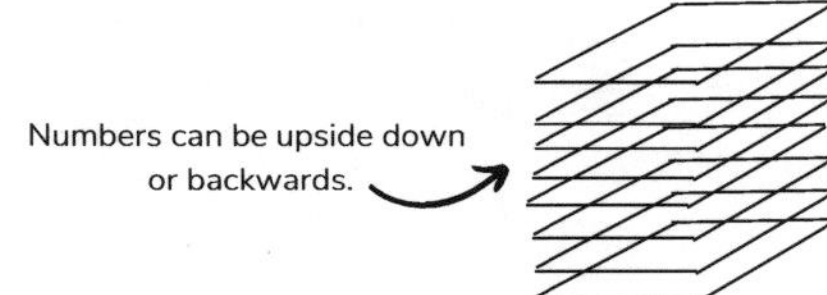

 | I can demonstrate an awareness and understanding of various emotions.

EMOJI EMOTIONS

Draw an emoji for each situation.

You earned some money.

You got a new game.

You don't think you did well on a test

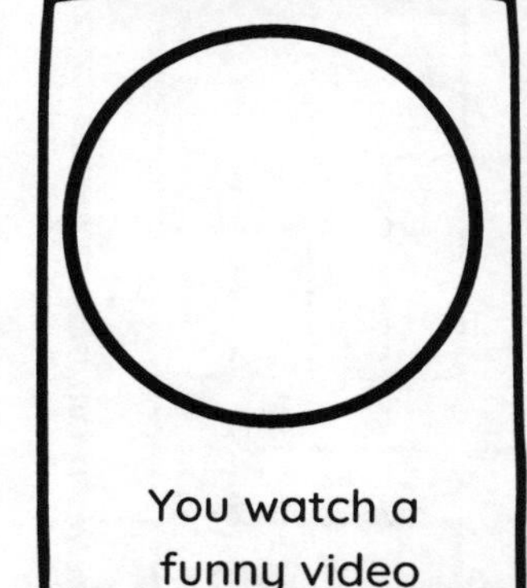

You watch a funny video

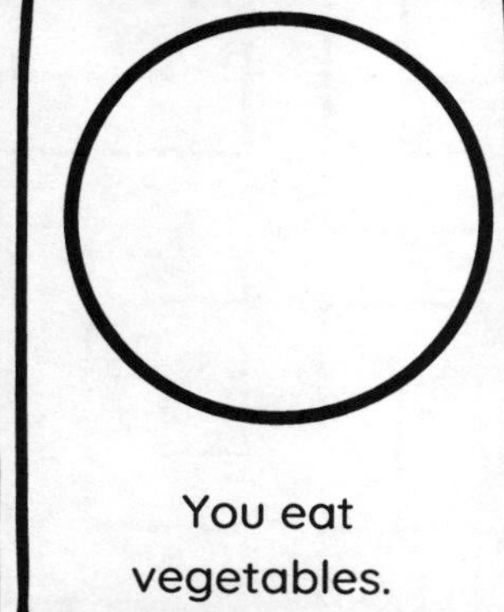

You eat vegetables.

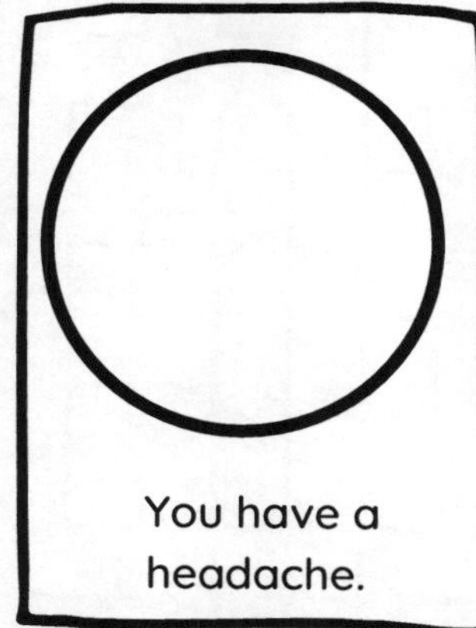

You have a headache.

Circle two words from the list to describe how you feel right now? Can't find your emotions in this list ? Feel free to write two other words below.

I feel this way because....

EMOTIONS WORD LIST

angry
annoyed
anxious
ashamed
awkward
brave
calm
cheerful
chill
confused
discouraged
disgusted
distracted
embarrassed
excited
friendly
guilty
happy
hopeful
jealous
joyful
lonely
loved
nervous
offended
scared
surprised
thoughtful
tired
uncomfortable
unsure
worried

Activity 32
K-4 PROBLEM SOLVING

I can systematically solve a problem by finding the route out of the maze.
I can use what I know to solve new problems.

Activity 33
1-4 MINDFUL COLORING

I can focus on how to choose and apply color in a design to bring an awareness to the present moment.

Activity 34
1-4 SOCIAL EMOTIONAL LEARNING

I can communicate an awareness of my own personal strengths.

Positive Self Talk

is important.

Trace your hand, on each finger

write or tell someone

something you like about yourself

or something you are good at.

Use this page to trace your hand.

On each finger, write something you like about yourself or something you are good at.

I can apply a range of communication and social skills to interact and communicate effectively.

THE POWER OF POSITIVE THINKING

Fill in the clouds...
When I am upset, these are five things that can make me feel better.

THERE IS ALWAYS MORE THAN ONE WAY TO DO SOMETHING

Activity 36
1-4 GEOMETRY

I can reason with shapes and their attributes.
I can explain my thinking and try to understand others.

WHICH ONE DOES NOT BELONG?

Choose someone special to share and discuss why the image you picked does not belong. Can you find more than one correct answer?

Build on what you know

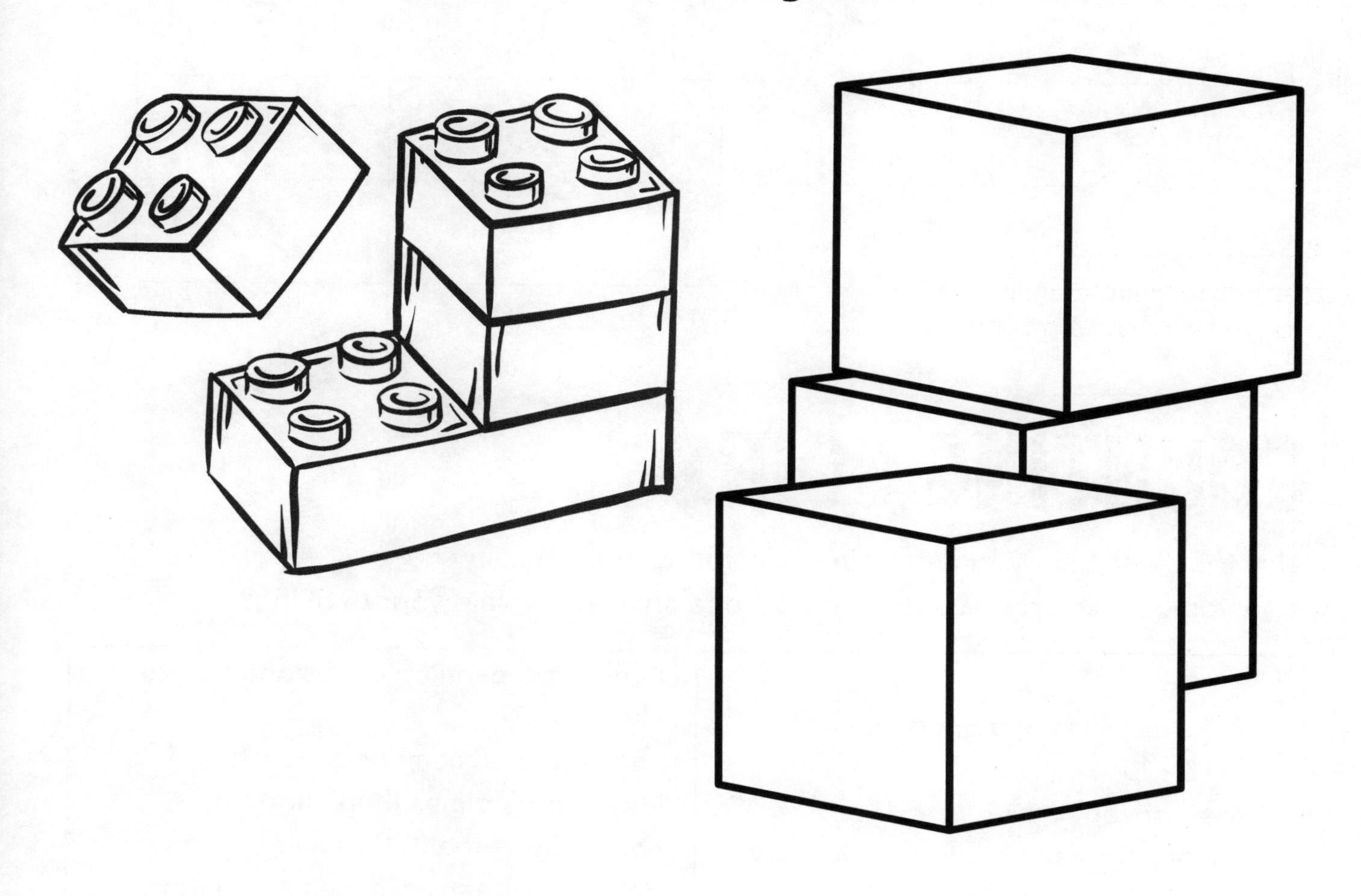

Activity 37
1-4 GEOMETRY
MEASUREMENT & DATA

I can work with money.
I can reason with attributes.
I can create a table to organize my thinking to systemically solve problems.

COUNTING CHANGE

Choose to think thoughts that make you feel good. Choose happy thoughts.

There are 18 ways you can have 32 cents. One way is listed in the chart.
How many ways can you make 32 cents? List the ways you can find in a similar chart.

25¢	10¢	5¢	1¢
1	0	0	7

COUNTING COINS

Use the clues below to determine how many of each coin I have.
Can you use pictures, words, symbols or a chart to explain your thinking?

I only have dimes and quarters.
I have $2.10
I have a dozen coins.
How many dimes and quarters do I have?

I only have pennies, dimes and quarters.
I have $1.29
I have an odd number of quarters.
I have more dimes than quarters.
I have fewer than 20 pennies.
How many pennies, dimes and quarters could I have?

Activity 38
1-4 GEOMETRY
MEASUREMENT & DATA

I can reason with shapes and their attributes.
I can understand concepts of area and recognize perimeter.
I can solve simple put-together, take-apart, and compare problems using information presented as a picture.

HOW MANY SQUARES CAN YOU FIND?

USE A DIFFERENT COLORED MARKER TO OUTLINE THE DIFFERENT SIZE SQUARES. RECORD HOW MANY SQUARES YOU CAN FIND OF EACH DIMENSION.	QUANTITY
1 X 1	
2 X 2	
3 X 3	
4 X 4	
TOTAL	

HOW MANY TRIANGLES CAN YOU FIND?

Want to Relax

TO CALM YOUR BRAIN, JUST BREATHE IN AND OUT DEEPLY, & COUNT TO FOUR

Activity 39
DAY DREAM

Activity 40
2-4 MEASUREMENT AND DATA

I can recognize, understand and measure area by counting unit squares.
I can find the perimeter of polygons with the same area and different perimeters.

TETROMINOS

A tetromino is a shape made of four squares, where each one touches the next along an entire side.

Can you find the shapes that mirror one another?
Color the shapes that mirror one another the same color.

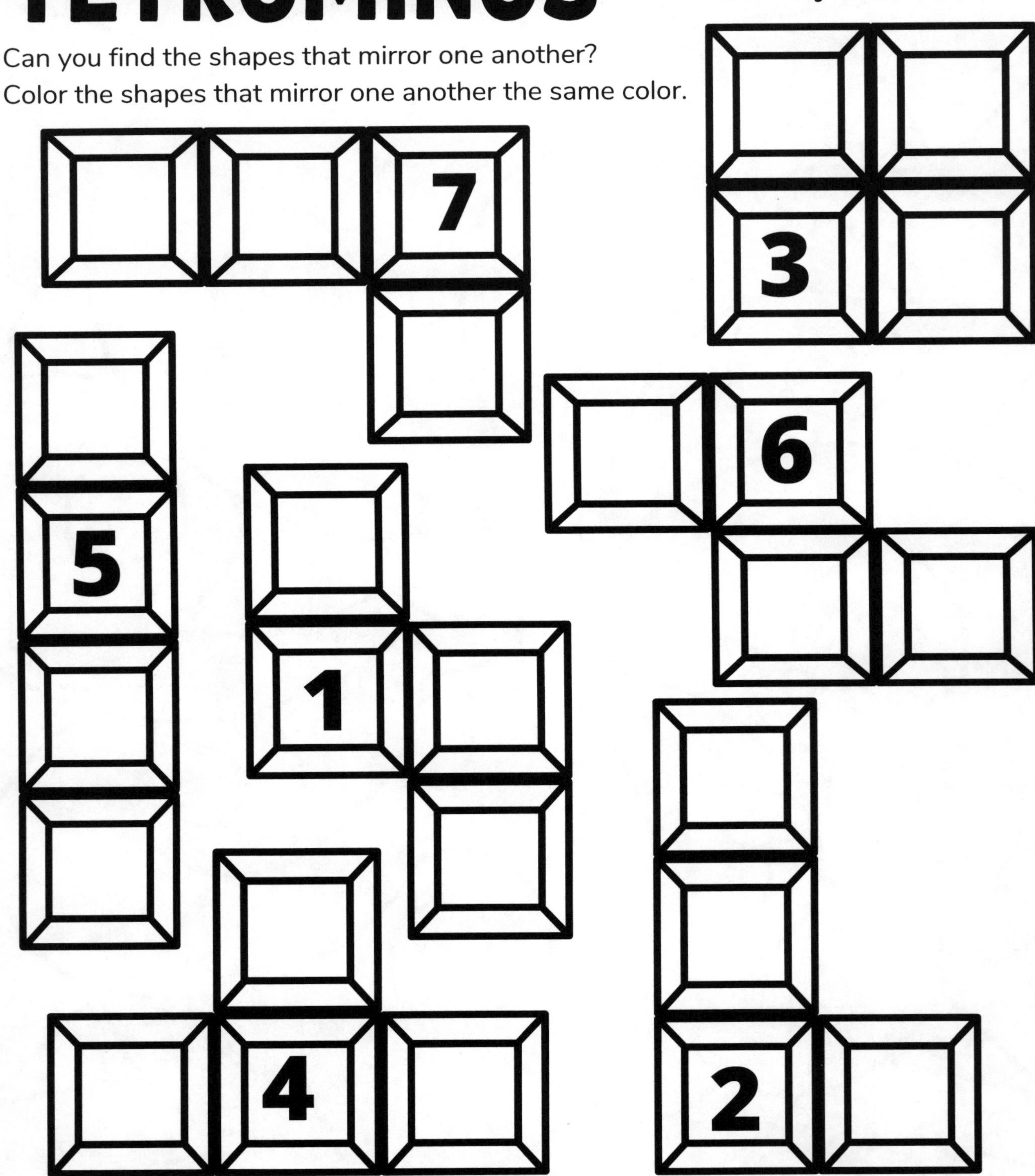

Activity 40
2-4 MEASUREMENT AND DATA

I can recognize, understand and measure area by counting unit squares.
I can find the perimeter of polygons with the same area and different perimeters.

TETROMINOS

Set One

Cut out the seven tetromino shapes out on the next two pages in order to try to build a few rectangles.

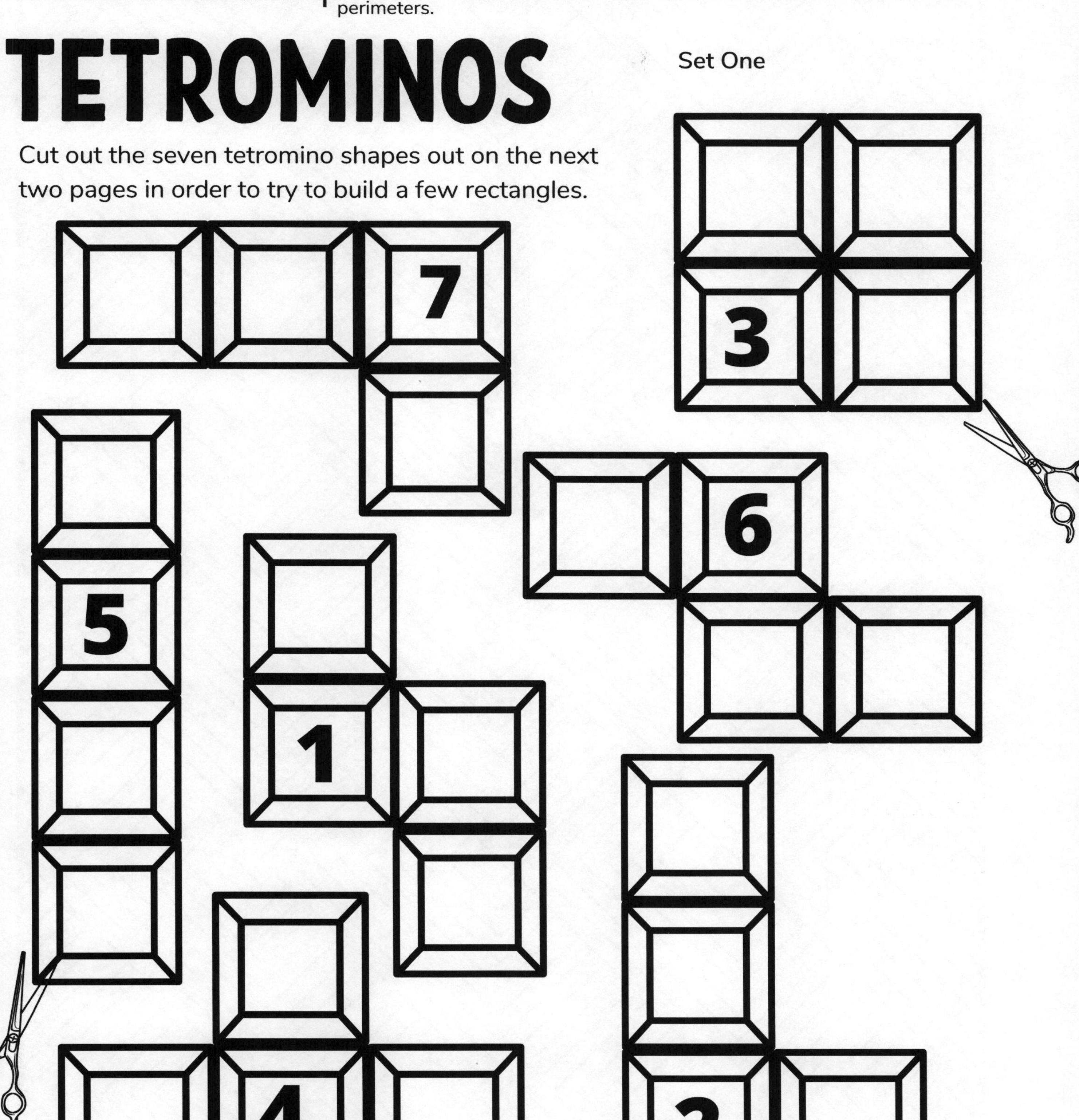

Activity 40
2-4 MEASUREMENT AND DATA

I can recognize, understand and measure area by counting unit squares.
I can find the perimeter of polygons with the same area and different perimeters.

TETROMINOS

Set Two

Cut out the seven tetromino shapes on the next two pages to explore complete

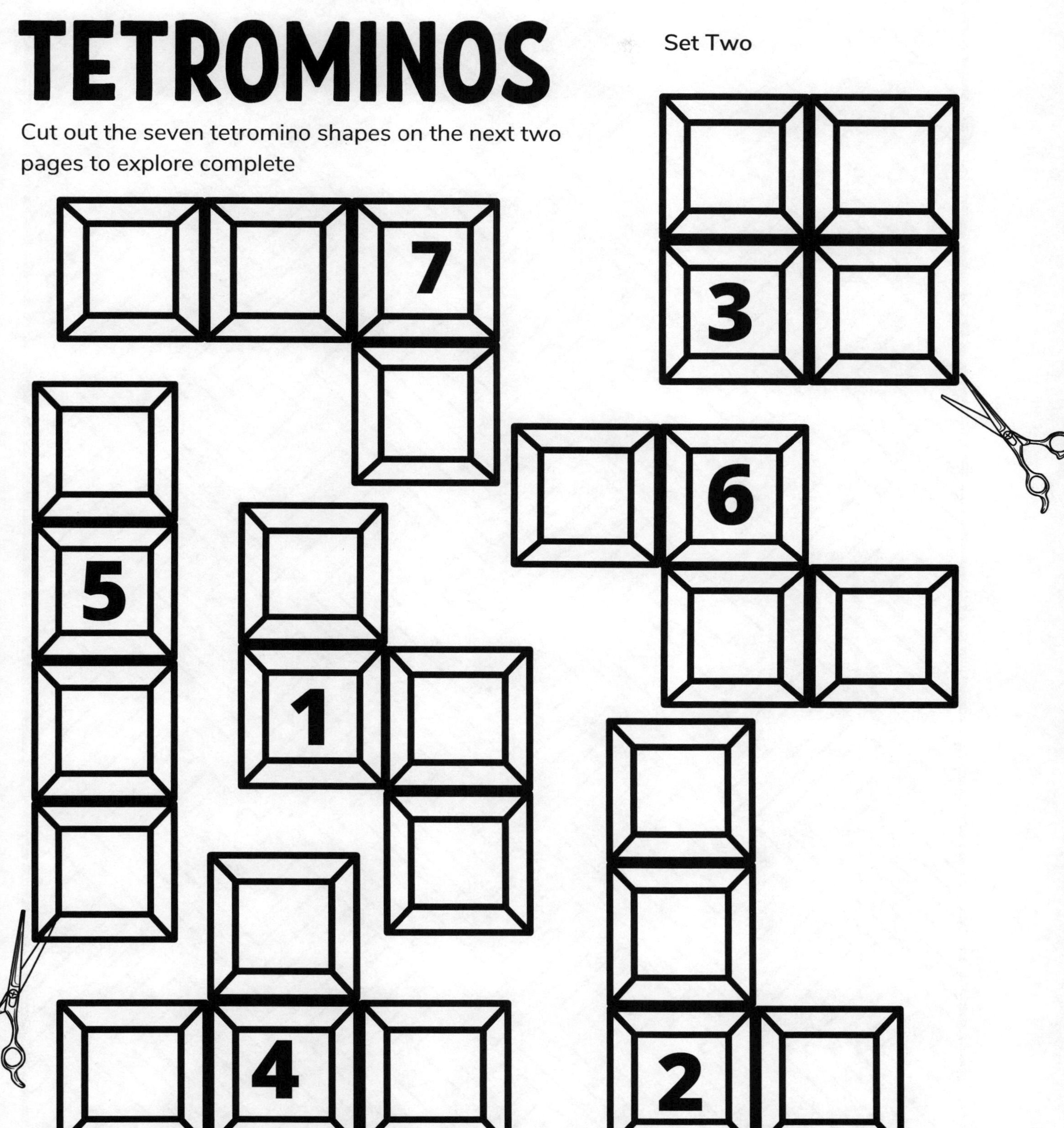

Activity 40
2-4 MEASUREMENT AND DATA | I can recognize, understand and measure area by counting unit squares. I can find the perimeter of polygons with the same area and different perimeters.

TETROMINOS

Can you make a rectangle using the 3 shapes described below?

Shape	Shape	Shape	YES? or NO?
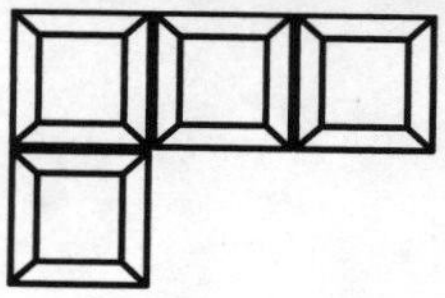	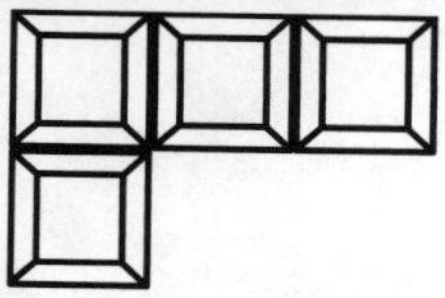	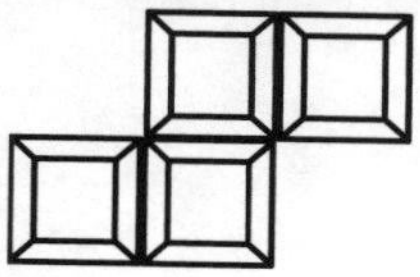	
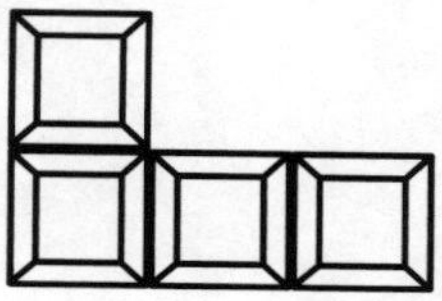	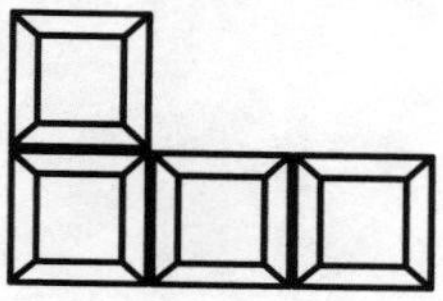	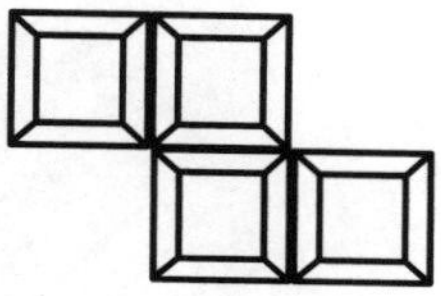	
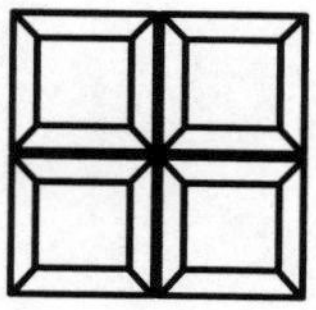	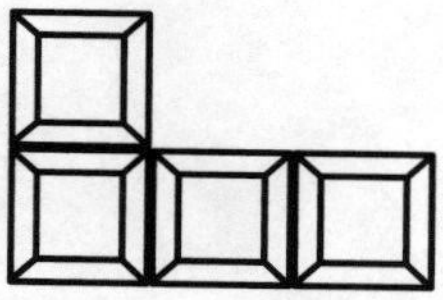	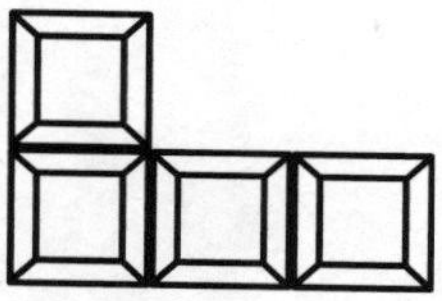	
	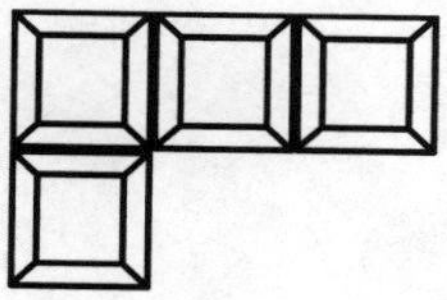	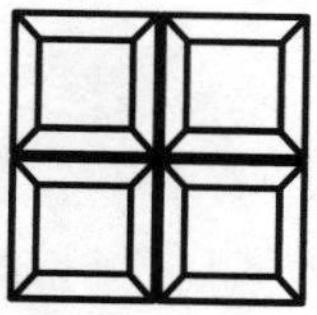	
	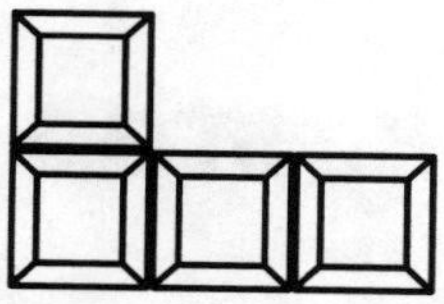	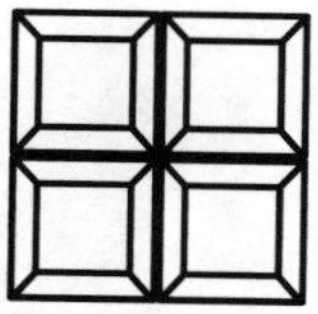	

IT'S OKAY

to take a walk, step way, and give yourself a break.

I can recognize, understand and measure area by counting unit squares.
I can find the perimeter and area of polygons.

TETROMINOS

Complete the table below. Find the tetromino with the largest perimeter.
Perimeter is the distance walking around the outside of the shape. What do you notice?

#	Shape	Area of Shape	Perimeter of Shape
1		4	10
2			
3			
4			
5			
6			
7			

You are special the way you are

Activity 41
2-4 MEASUREMENT AND DATA | I can apply compass directions to a map.

N
NW NE
W E
SW SE
S

COMPASS DIRECTIONS

Let Self-Love Be Your Compass

allow space for self-acceptance, take steps that support your self-worth.

1. From the start, go east 2 squares. Where are you now?
2. Go South-West one square. Where are you now?
3. Go West 2 squares. Where are you now?
4. Go North 2 squares. Where are you now?
5. Start at the bus stop, how do you get to the cafe?
6. Write directions for someone to follow from anywhere on the map to a different place on the map.

Activity 42
1-4 SOCIAL EMOTIONAL LEARNING

I can spend 30 minutes a day to enhance my health and well being.

WHAT HAVE YOU DONE FOR YOURSELF TODAY?

1. ____________________

2. ____________________

3. ____________________

I can understand and show whether or not a 2-dimensional shape can cover or tile a region.

TESSELLATE

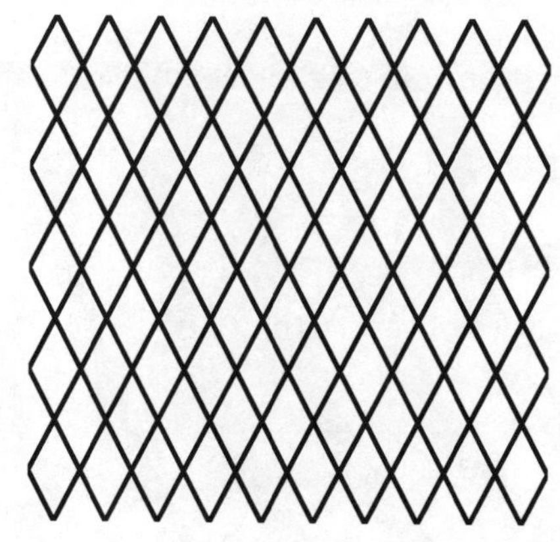

To tessellate a plane, means to cover or tile a region, such as the rectangle at the left with a given shape leaving no gaps, holes or empty white spaces.

Cut out the shapes below, try tracing one shape at a time to cover or tile the rectangle at the left. If you prefer you could try tiling a whole sheet of paper. Identify which shape below cannot tessellate or cover the rectangular region.

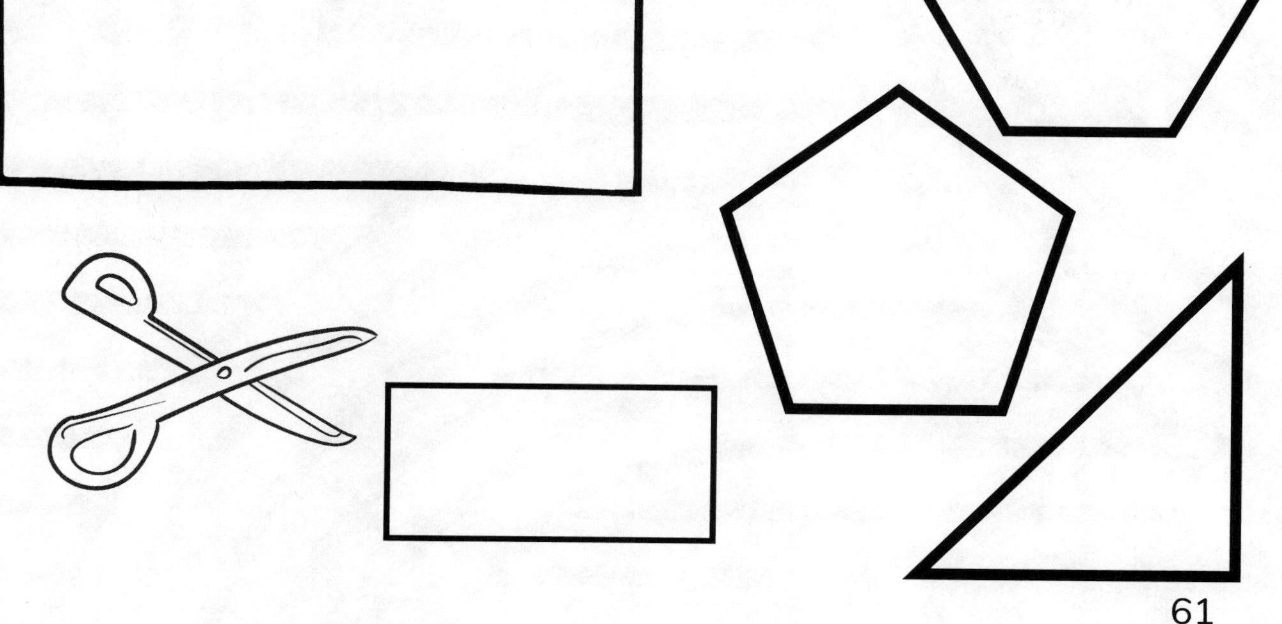

I can focus on how to choose and apply color in a design to bring an awareness to the present moment.

I can use addition and multiplication to solve problems.
I can solve problems without giving up.

DIAMOND PROBLEMS

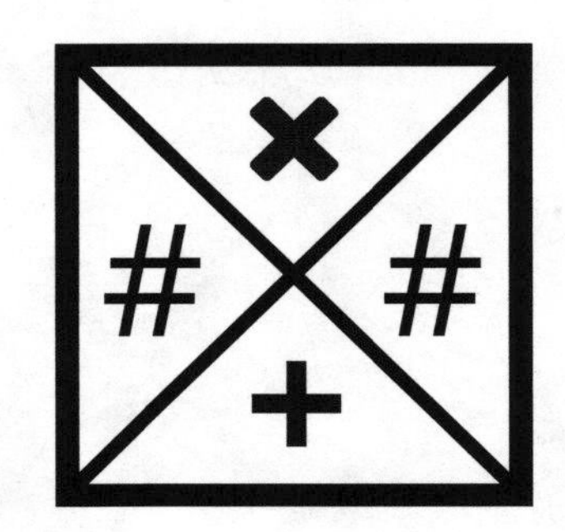

The top cell contains the **product** of the two numbers in the left and right cells and the bottom cell contains the **sum** of the left and right cells.

A **product** is the answer to a multiplication problem, and a **sum** is the answer to an addition problem.

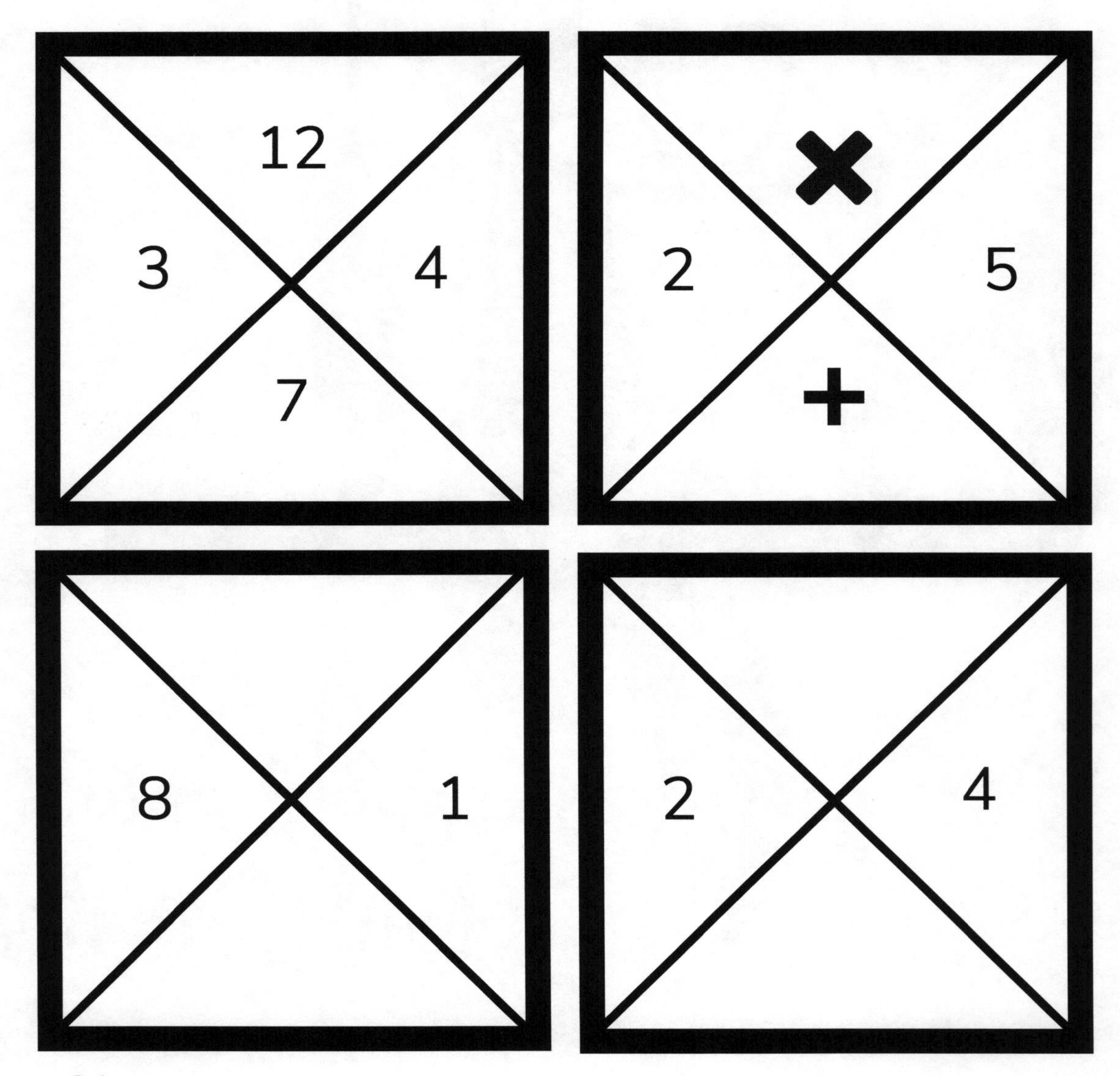

I can use addition and multiplication to solve problems.
I can solve problems without giving up.

DIAMOND PROBLEMS

The top cell contains the **product** of the two numbers in the left and right cells and the bottom cell contains the **sum** of the left and right cells.

A **product** is the answer to a multiplication problem, and a **sum** is the answer to an addition problem.

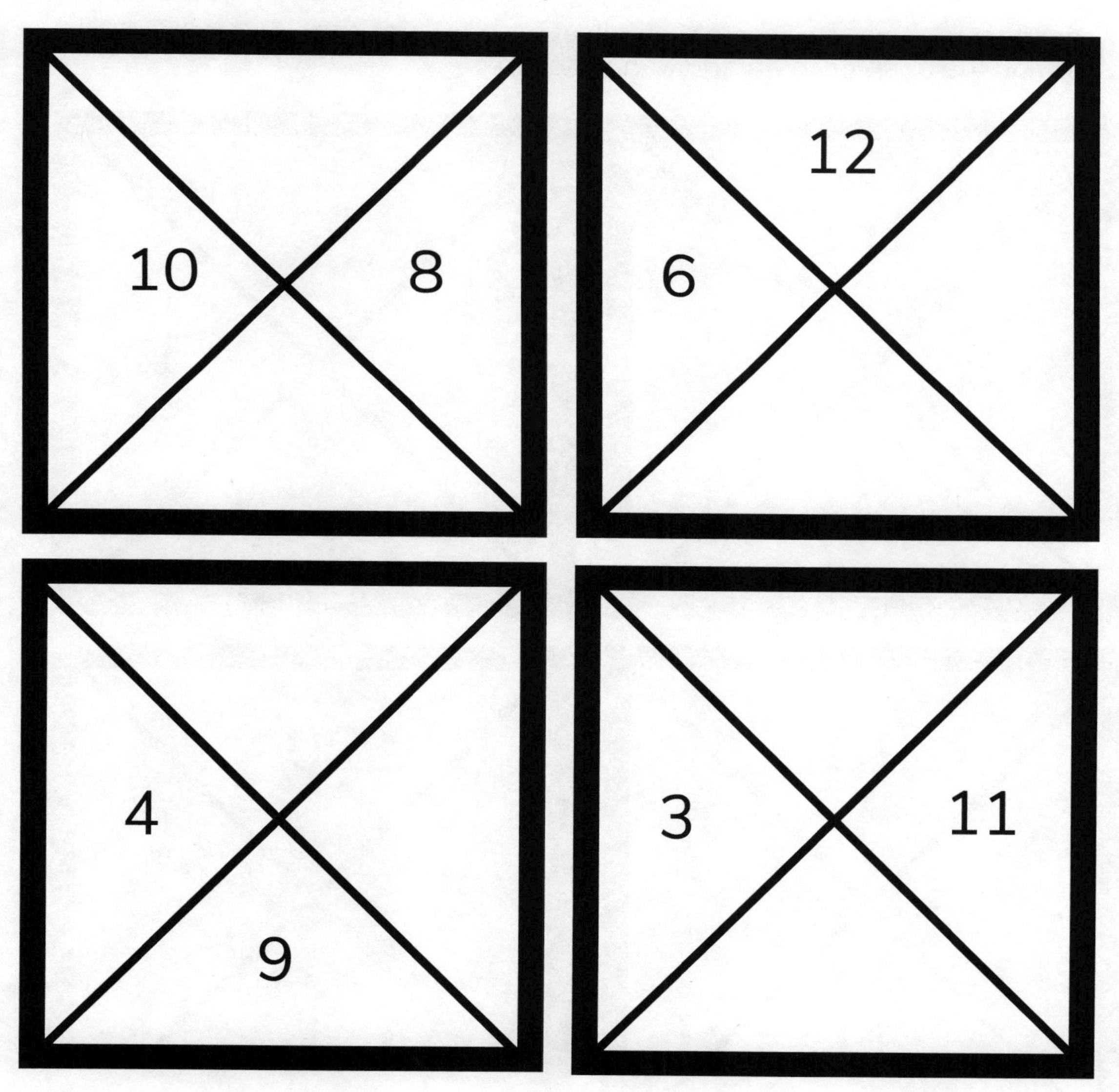

DIAMOND PROBLEMS

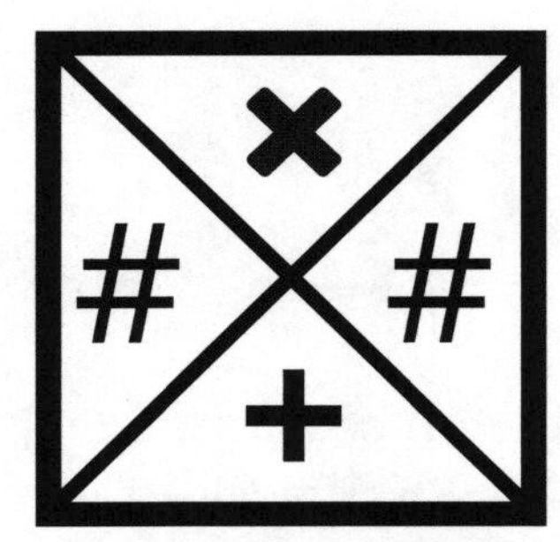

The top cell contains the **product** of the two numbers in the left and right cells and the bottom cell contains the **sum** of the left and right cells.

A **product** is the answer to a multiplication problem, and a **sum** is the answer to an addition problem.

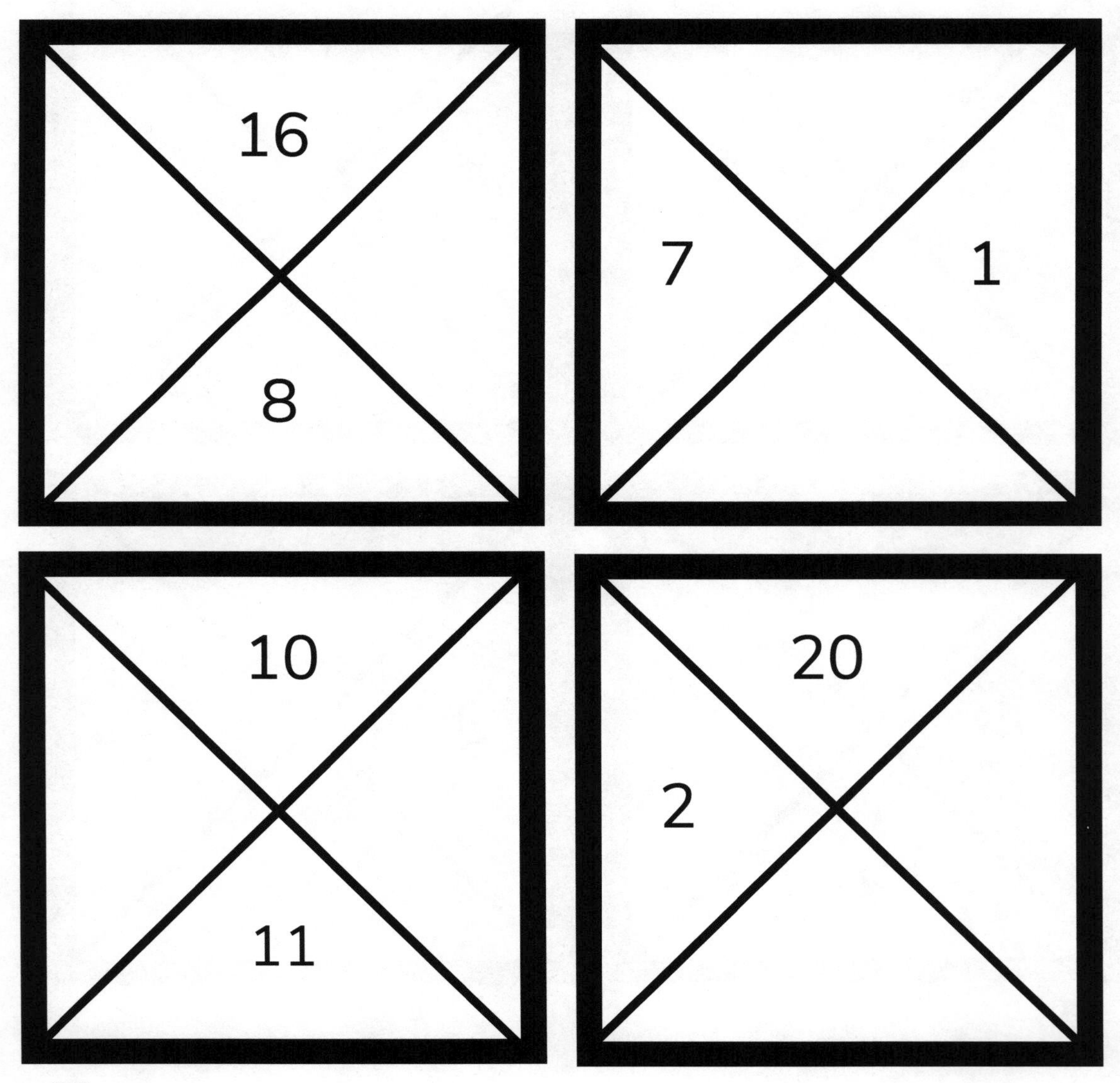

Activity 45
1-4 OPERATIONS AND ALGEBRAIC THINKING

I can use addition and multiplication to solve problems.
I can solve problems without giving up.

DIAMOND PROBLEMS

The top cell contains the **product** of the two numbers in the left and right cells and the bottom cell contains the **sum** of the left and right cells.

A **product** is the answer to a multiplication problem, and a **sum** is the answer to an addition problem.

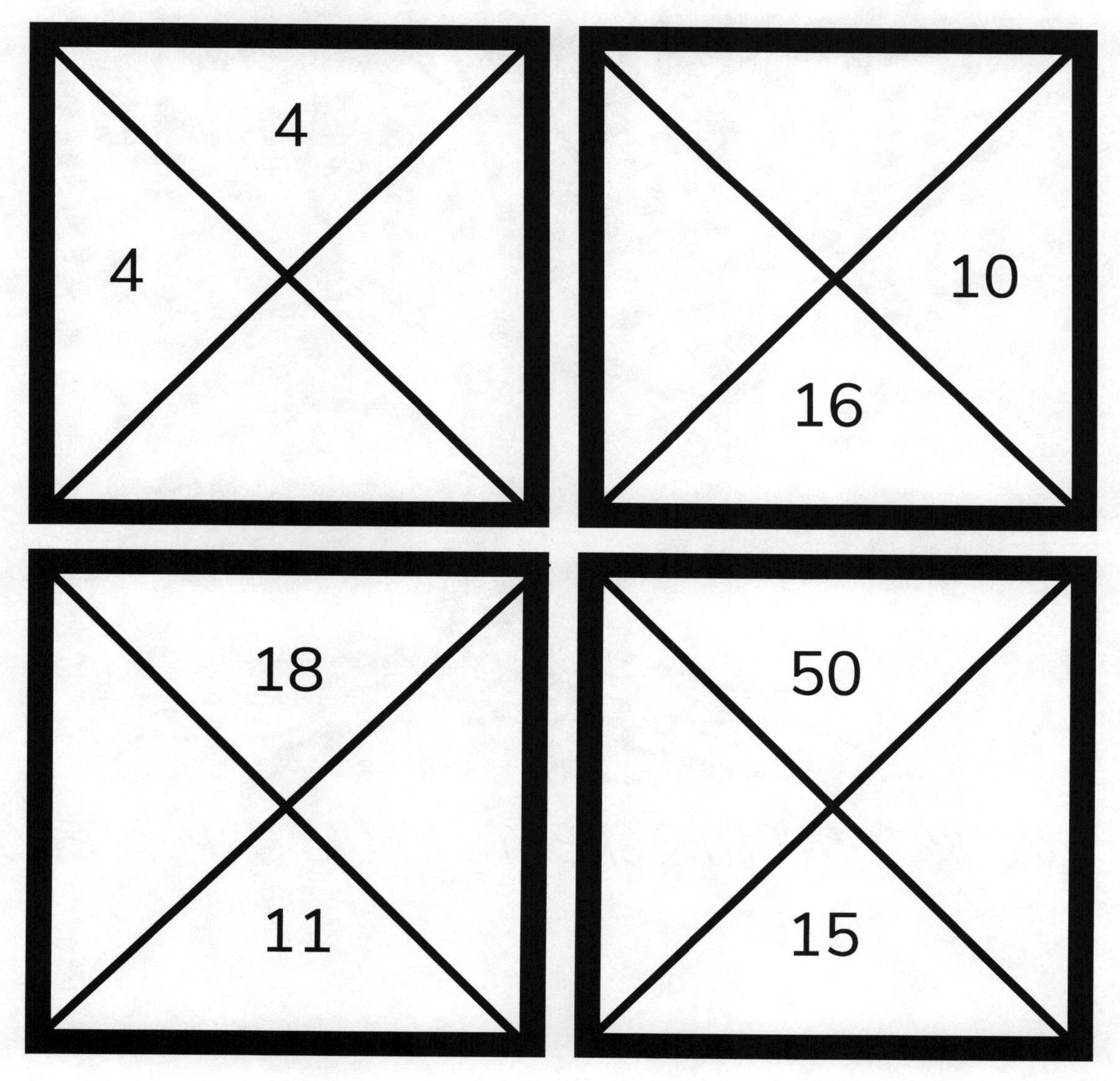

YOU ARE ALLOWED TO CHANGE FOR THE BETTER

YOUR THOUGHTS CREATE YOUR EXPERIENCES

HAVE PATIENCE
STAY SILENT

Activity 47
2-4 OPERATIONS AND ALGEBRAIC THINKING

I can use the four operations to solve problems.
I can show my work in many ways.

SKETCH & SOLVE

A starfish has five arms. There are six starfish on the beach. How many starfish arms are on the beach? ?

Each horse has four legs, and five horses are in the barn. Two horses leave the barn. How many horse legs are left in the barn?

Activity 47
2-4 OPERATIONS AND ALGEBRAIC THINKING

I can use the four operations to solve problems.
I can show my work in many ways.

SKETCH & SOLVE

A starfish has five arms. There are six starfish on the beach. How many starfish arms are on the beach? ?

Each horse has four legs, and five horses are in the barn. Two horses leave the barn. How many horse legs are left in the barn?

I can use the four operations to solve problems.
I can show my work in many ways.

SKETCH & SOLVE

Twenty three penquins were on the ice. How many feet were stepping on the ice?

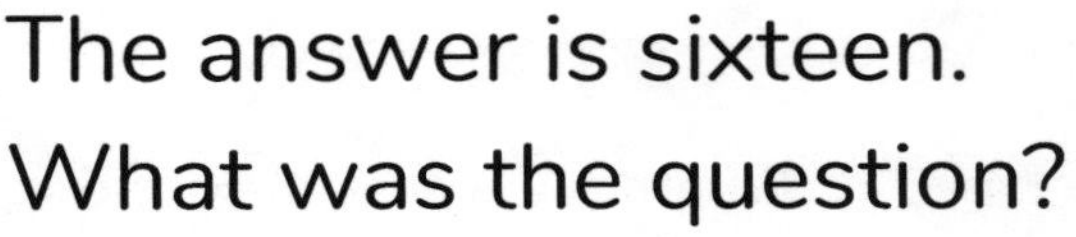

The answer is sixteen. What was the question?

Activity 47
2-4 OPERATIONS AND ALGEBRAIC THINKING

I can use the four operations to solve problems.
I can show my work in many ways.

SKETCH & SOLVE

A beetle has six legs. Six beetles are in the garden. How many beetle legs are in the garden?

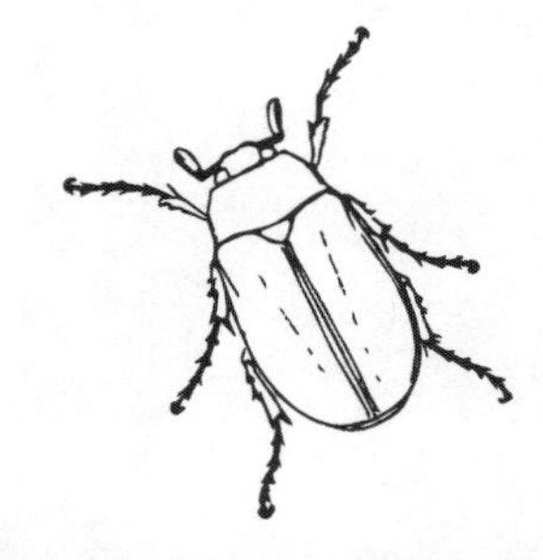

Imagine seeing nine zebras riding bicycles down the street. What could be the question?

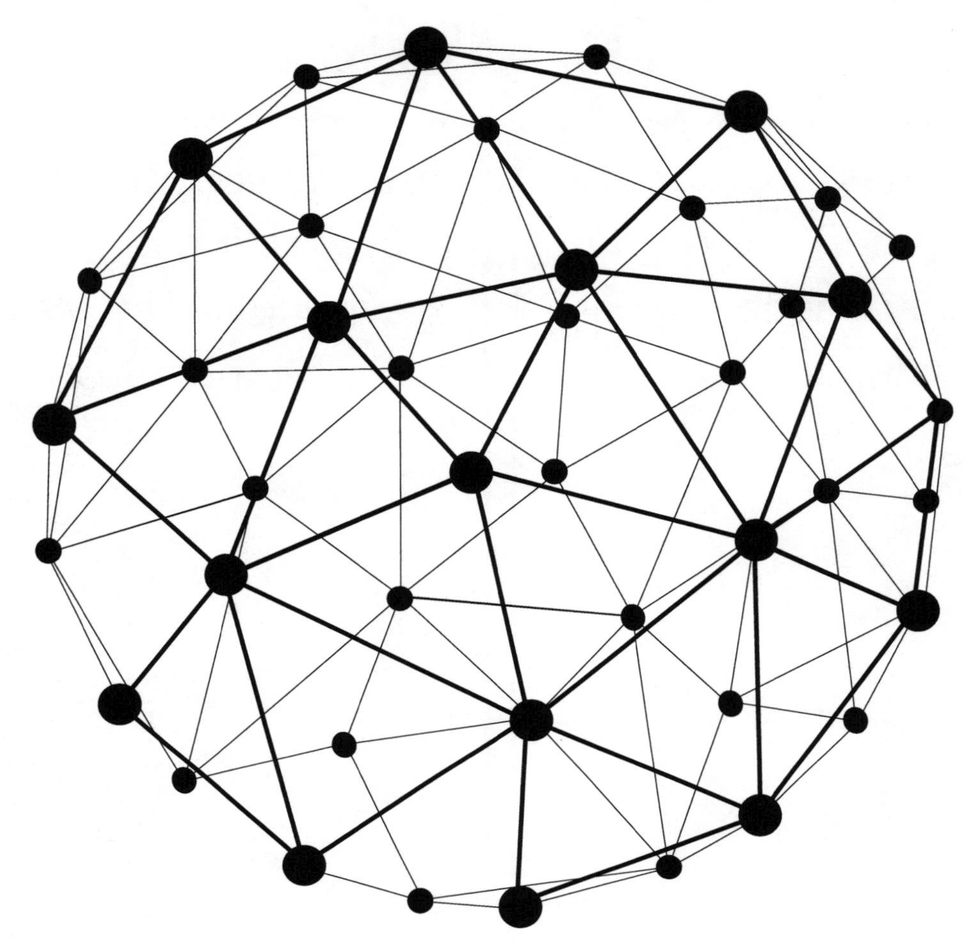

Step out of your comfort zone

Connect with others

Make new friends

I can focus on how to choose and apply color in a design to bring an awareness to the present moment.

Let Go
OF THE
THINGS
YOU CAN'T
ctrl

Activity 49
1-4 SOCIAL EMOTIONAL LEARNING

I can demonstrate the skills to express my emotions.
I can apply a range of communication and social skills to interact and communicate effectively.

What are some anxious thoughts or worries that you might need to let go? Write them on the kites above, then talk about these thoughts with someone who cares about you.

Activity 50
1-4 PROBLEM-SOLVING

I can systematically solve a problem by finding the route out of the maze.
I can use what I know to solve new problems.

YOU GOT THIS!

Activity 51
1-4 PROBLEM SOLVING

I can use what I know to solve new problems.
I can solve problems without giving up.

FOUR COLOR PROBLEM

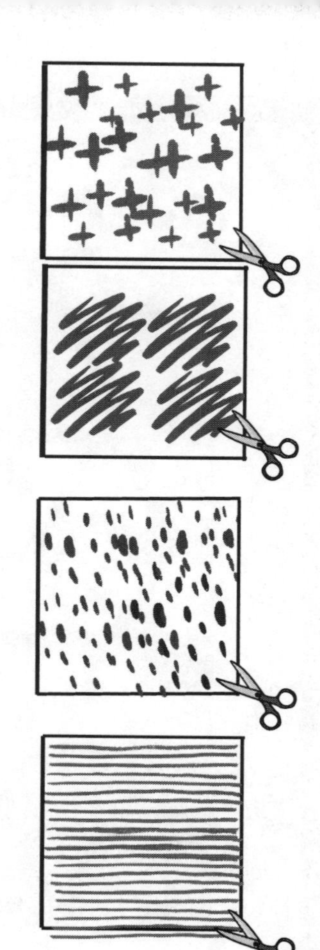

The Four Color Theorem states that no more than four colors are required to color the regions of any map so that no two adjacent regions have the same color. Regions that meet only at a corner may be the same color.
This was the first major theorem proved using a computer!

Cut or tear off little bits of the pattern tiles and place them on the map so no two adjacent regions have the same pattern. Then using only four different colored markers, color the regions to show your final answer.

Activity 51
1-4 PROBLEM SOLVING

I can use what I know to solve new problems.
I can solve problems without giving up.

FOUR COLOR PROBLEM

The Four Color Theorem states that no more than four colors are required to color the regions of any map so that no two adjacent regions have the same color. Regions that meet only at a corner may be the same color.
This was the first major theorem proved using a computer!

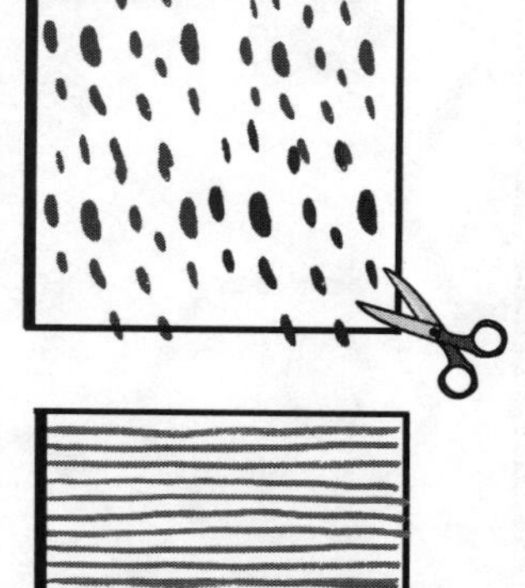

Cut or tear off little bits of the pattern tiles and place them on the map so no two adjacent regions have the same pattern. Then using only four different colored markers, color the regions to show your final answer.

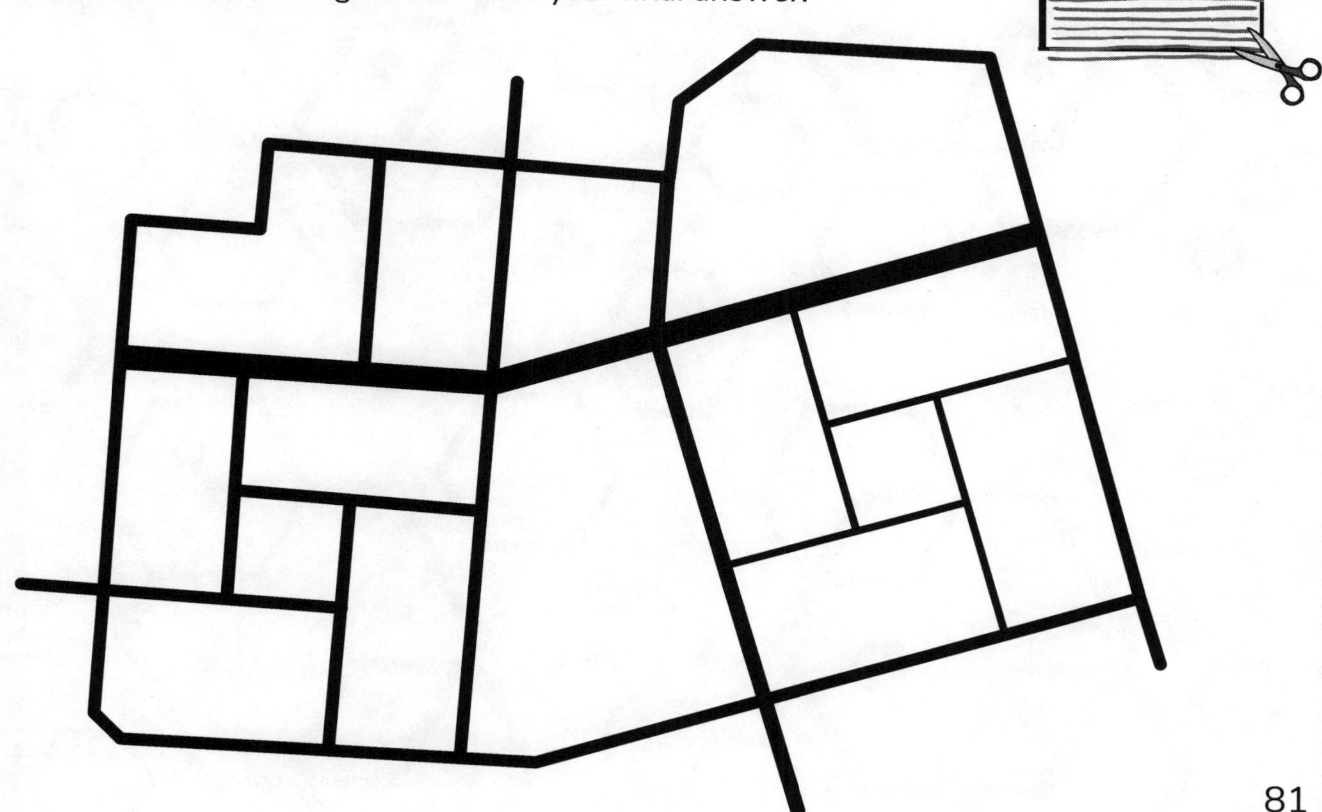

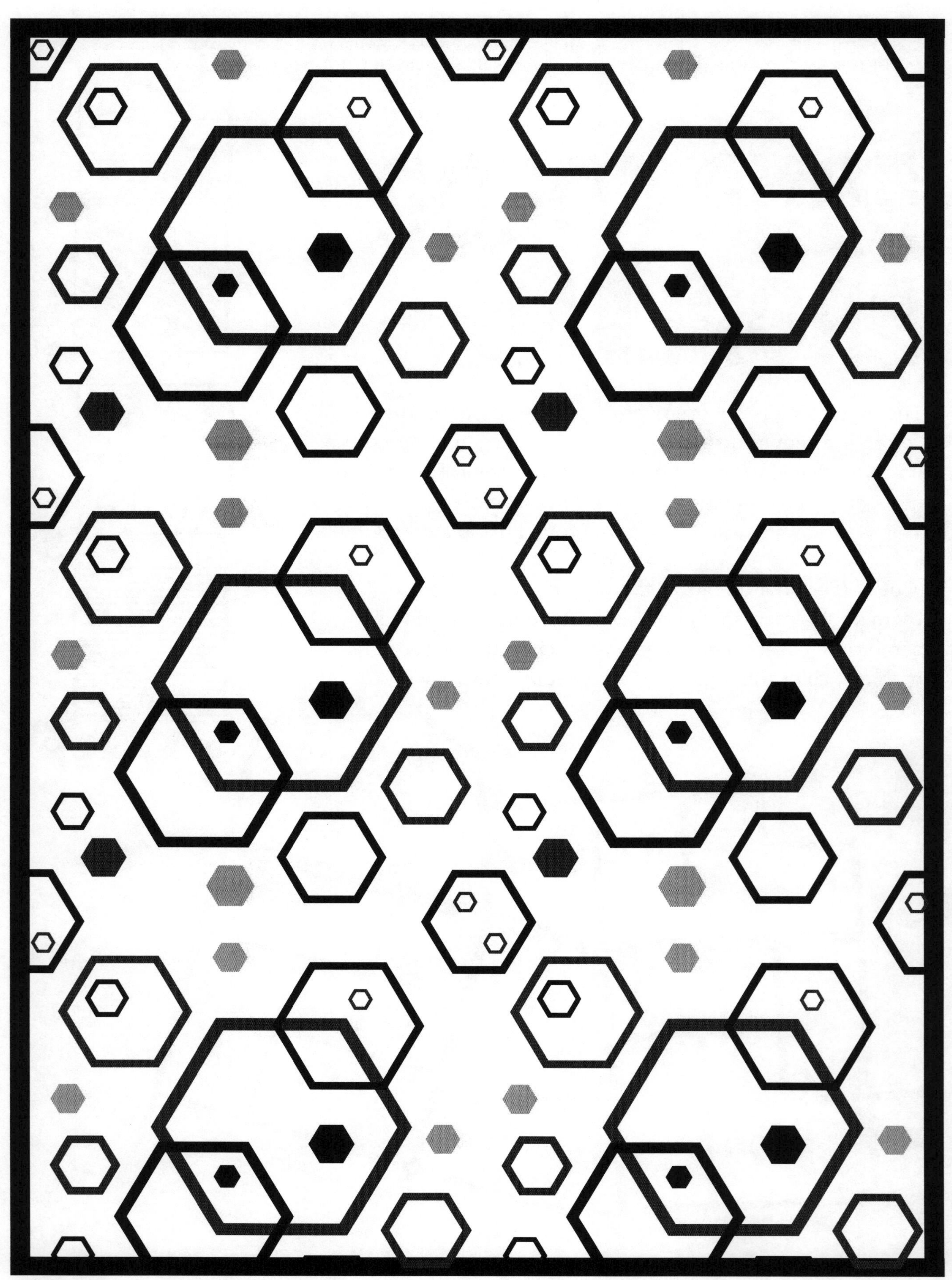

I can use what I know to solve new problems.
I can solve problems without giving up.

FOUR COLOR PROBLEM

Color the design below using only four colors. No shape of the same color should touch an adjacent shape.

Write what happens in the next frame of this comic?

"If you want happiness for an hour, take a nap.
If you want happiness for a day, go fishing.
If you want happiness for a year, inherit a fortune.
If you want happiness for a lifetime, help somebody."
-Chinese Proverb

I can count by prime numbers.
I can work carefully and check my work.

WHAT IS A PRIME NUMBER?

A number is **prime** if it has only two different factors, one and itself.

The number 2 is a prime number because it has only two different factors, 1 and 2.

The number 4 is not a prime number because it has three different factors, 1, 2 and 4.

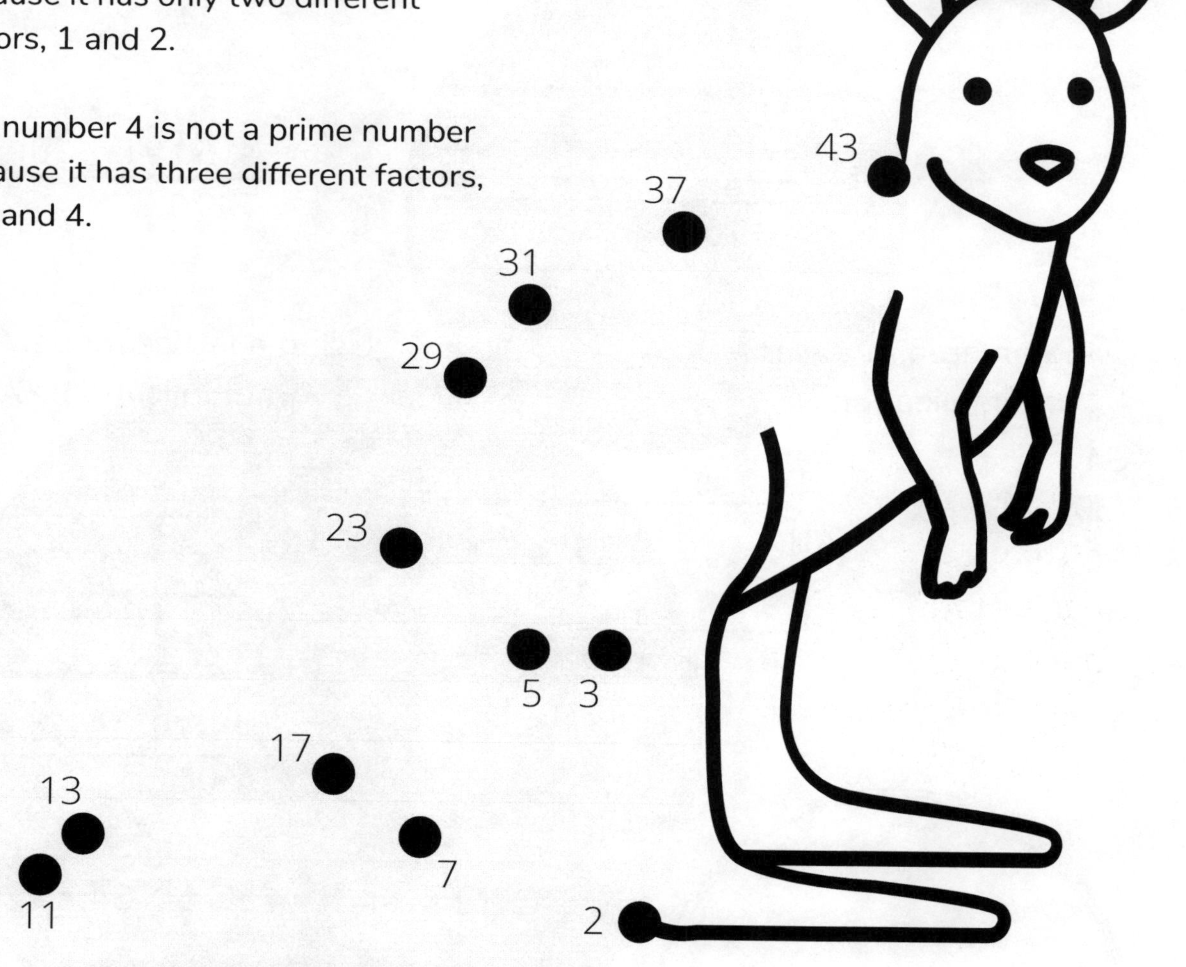

What do you notice if you find the sum of any two prime numbers?
What do you notice if you find the sum of any three prime numbers?

Activity 53
3-4 OPERATIONS AND ALGEBRAIC THINKING

I can use place value understanding and an area model to multiply a whole number with two digits by a whole number with one digit.
I can show my work in many ways.

FOUR WAYS TO MULTIPLY

Study and reason through the four examples:

#1 Understanding multiplication is repeated addition:

5 X 34 =

$$\begin{array}{r} 34 \\ 34 \\ 34 \\ 34 \\ +\ 34 \\ \hline 170 \end{array}$$

#2 Using place value for multiplication:

TENS ONES

$$\begin{array}{rl} 34 & \\ \times\ 5 & \\ \hline 20 & = 5 \times 4 \\ +150 & = 5 \times 30 \\ \hline 170 & \end{array}$$

#3 Using an area model for multiplication:

5 X 34 =
5 X (30 + 4) =
5 X 30 + 5 X 4 =
150 + 20 = 170
5 X 34 = 170

	30	4
5	150	20

Total Area = 170

#4 Using a lattice model for multiplication: 5 X 34 =

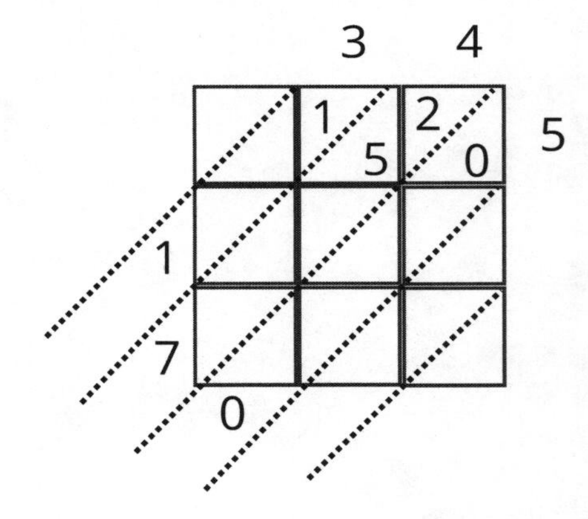

5 X 34 = 170

#1 Draw the lattice.
#2 Label the sides.
#3 Multiply the numbers.
#4 Add down the diagonals.
#5 Write your answer, reading from the top left to the bottom right.

Your turn:

Find 62 X 3 , by the method of your choice.

Activity 54
3-4 OPERATIONS AND ALGEBRAIC THINKING

I can use place value understanding and an area model to multiply a whole number with two digits by a whole number with two digits.
I can show my work in many ways.

PRACTICE MULTIPLYING

PLACE VALUE METHOD

$$\begin{array}{r} 12 \\ \times\ 36 \\ \hline \end{array}$$

12 = 6 X 2
60 = 6 X 10
60 = 30 X 2
\+ 300 = 30 X 10

432

Your Turn

$$\begin{array}{r} 25 \\ \times\ 41 \\ \hline \end{array}$$

AREA MODEL METHOD

12 X 36 =

(10 + 2) X (30 + 6) =

300 + 60 + 60 +12 = 432

	30	6
10	300	60
2	60	12

Your Turn

25 X 41=

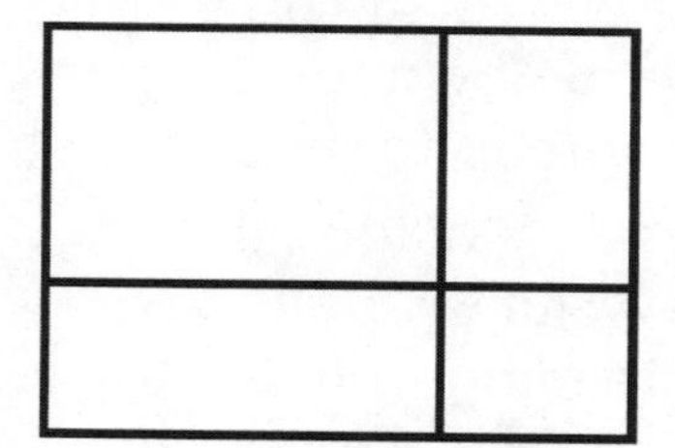

Activity 54
3-4 OPERATIONS AND ALGEBRAIC THINKING

I can use place value understanding and an area model to multiply a whole number with two digits by a whole number with two digits.
I can show my work in many ways.

PRACTICE MULTIPLYING

LATTICE METHOD

$$\begin{array}{r} 12 \\ \times\ 36 \\ \hline 432 \end{array}$$

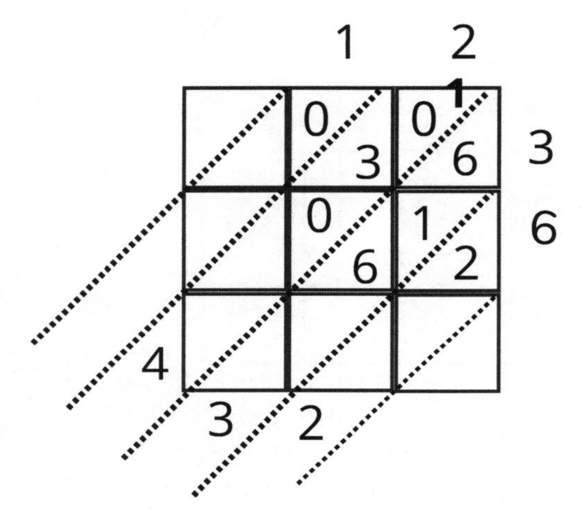

#1 Draw the lattice.
#2 Label the sides.
#3 Multiply the numbers.
#4 Add down the diagonals, regroup to the next left diagonal.
#5 Write your answer, reading from the top left to the bottom right.

LATTICE METHOD

Your Turn

$$\begin{array}{r} 25 \\ \times\ 41 \\ \hline \end{array}$$

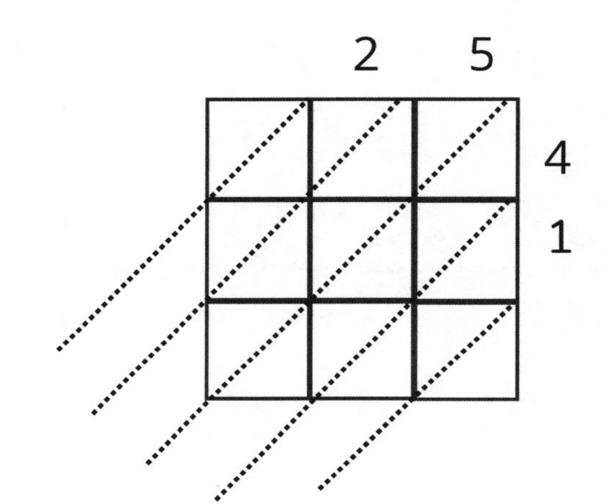

Activity 55
1-4 CREATIVE THINKING

I can hold a growth mindset and solve open ended problems.
I can solve problems without giving up.

IMAGINATION WORKOUT #2

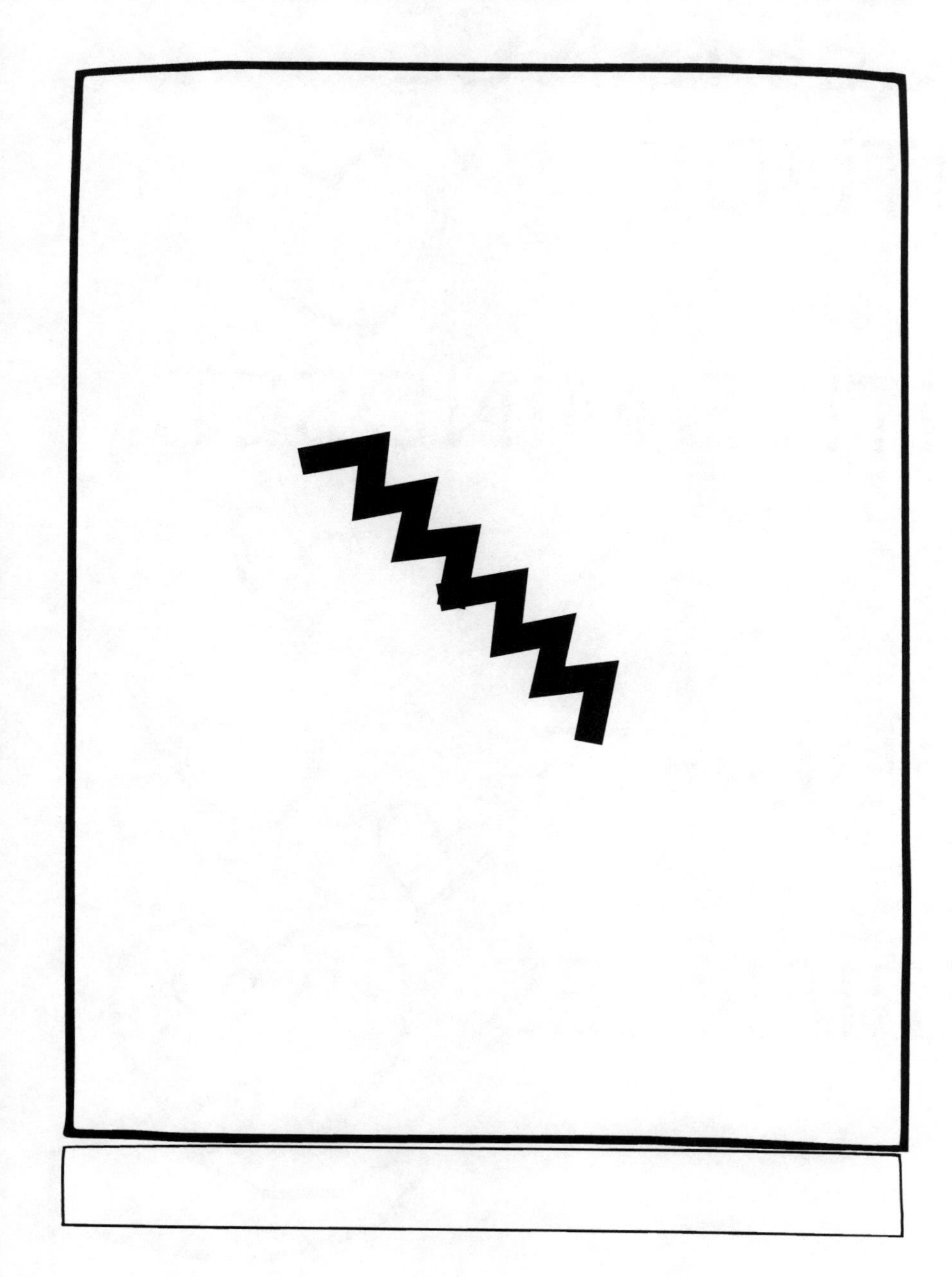

Imagine what the design could be, then finish the drawing yourself!
Give your drawing a title.

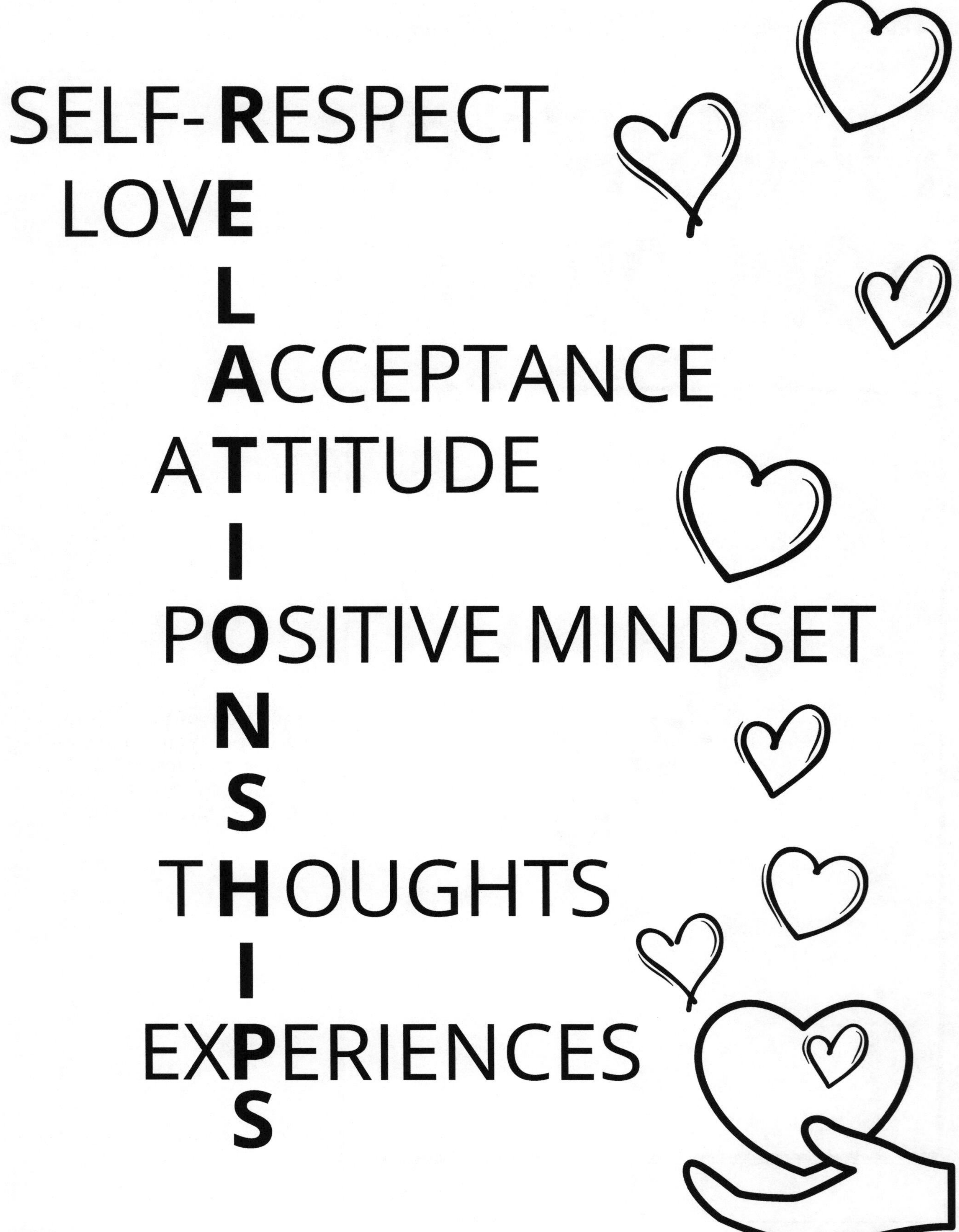
SELF-RESPECT
LOVE
L
ACCEPTANCE
ATTITUDE
I
POSITIVE MINDSET
N
S
THOUGHTS
I
EXPERIENCES
S

Activity 56
3-5 MINDFUL COLORING &
SOCIAL EMOTIONAL LEARNING

I can begin to demonstrate awareness of my personal rights, feelings and responsibilities.
I can focus on how to choose and apply color in a design to bring an awareness to the present moment.

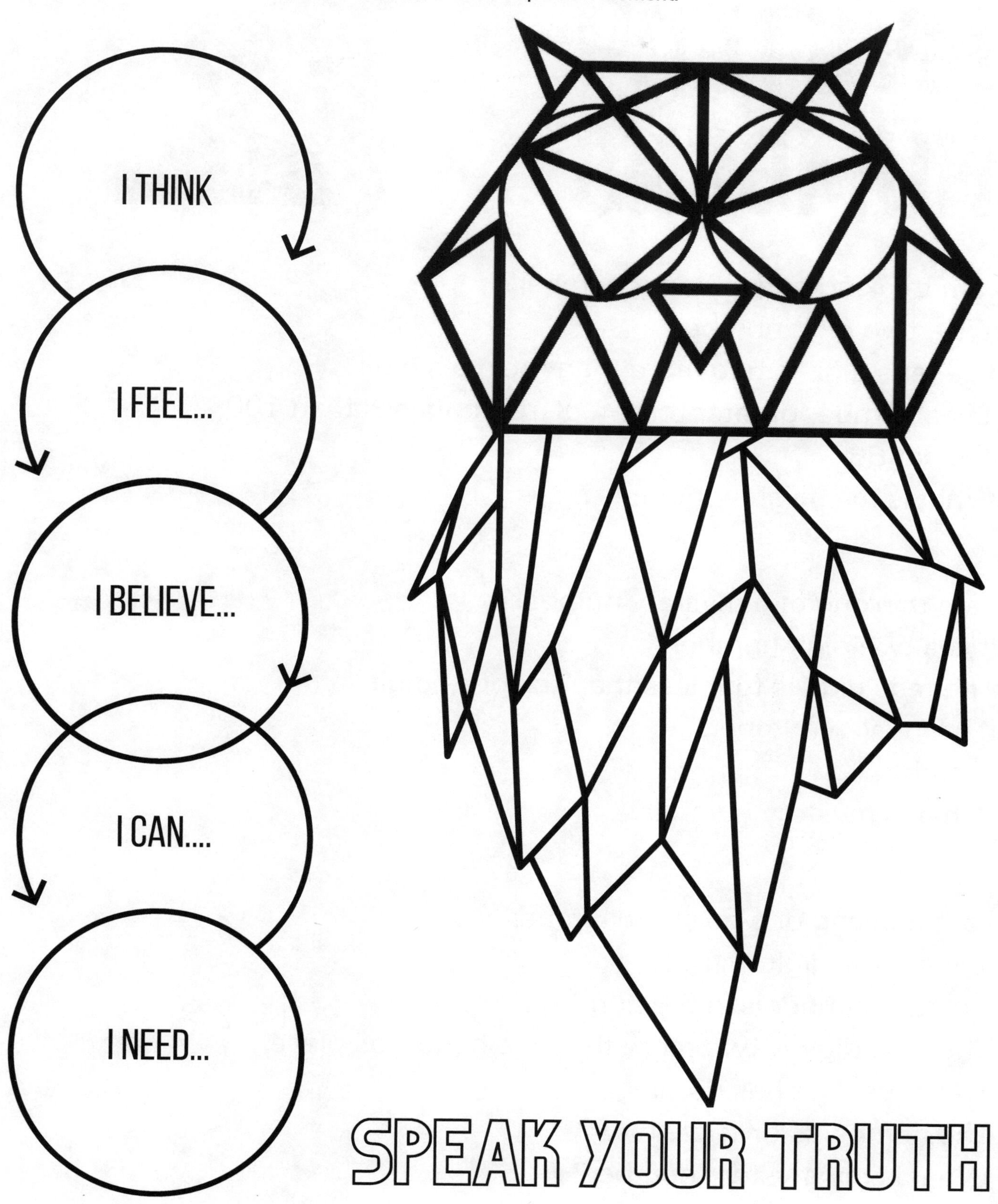

I can generalize an understanding of place value.
I can think about numbers in many ways.

MYSTERY NUMBERS

I am thinking of a mystery number?
It is a two digit number.
The tens digit is two more than the ones digit.
It is a number greater than $\frac{3}{4}$ of 100 and less than 100.
It is not 86.
What is my mystery number?

I am thinking of a mystery number?
It is a two digit number.
The tens digit is four less than the ones digit.
It is an even number.
It is not 48.
What is mystery number?

I am thinking of a mystery number?
It is a three digit number.
An odd number is in the hundreds place.
The tens digit is two more than the hundreds place.
The ones digit has nothing.
The number is divisible by 30.
What is my mystery number?

Activity 57
3-5 OPERATIONS AND ALGEBRAIC THINKING

I can use inverse operations to solve problems.
I can think about numbers in many ways.

MYSTERY NUMBERS

I am thinking of a mystery number?
I double the number and I get 26.
What was my mystery number?

I am thinking of a mystery number?
I add 15 and I get 35.
What was my mystery number?

I am thinking of a mystery number?
I take away 13 and I get 62.
What was my mystery number?

I am thinking of a mystery number?
I split the number in half then add 5 and I get 47.
What was my mystery number?

I am thinking of a mystery number?
I triple the number then subtract 4 and I get 11.
What was my mystery number?

Activity 58
3-5 OPERATIONS AND ALGEBRAIC THINKING

I can use inverse operations to solve problems.
I can understand palindrome numbers.
I can think about numbers in many ways.

PALINDROMIC NUMBERS

A **palindrome** number is a number that reads the same forwards or backwards. The number 101 is a palindrome. The number 21 is not a palindrome.

If you take any number and add its numbers reversed (or its inverse), the sum will eventually become a palindrome or a palindromic number.

6 } 6 is already a palindrome or 6 is a zero step palindrome

55 } 55 is a zero step palindrome

101 } 101 is a zero step palindrome

$$\begin{array}{r} 21 \\ +12 \\ \hline 33 \end{array}$$ 21 is a one step palindrome

$$\begin{array}{r} 93 \\ +39 \\ \hline 132 \\ +231 \\ \hline 363 \end{array}$$ 93 is a two step palindrome

89 and 98 } are 24 step palindromes!.

How many **palindrome** numbers can you find between 1 and 100? Use the hundred chart to help you keep track of the palindromes you can find.

1. How many are zero step palindromes? __________
2. How many are one step palindromes? __________
3. How many are two step palindromes? __________
4. How many are three step palindromes? __________
5. How many are four step palindromes? __________
6. How many are five step palindromes? __________
7. How many are six step palindromes? __________

What patterns do you see?

I can use inverse operations to solve problems.
I can think about numbers in many ways.

PALINDROMIC NUMBERS

Color the palindromic numbers:

RED= already a palindrome or a zero step palindrome
YELLOW= one step palindrome
BROWN= two step palindrome
GREEN= three step palindrome
BLUE= four step palindrome
ORANGE = five step palindrome
PURPLE= six step palindrome

1	2	3	4	5	6	7	8	9	10
11	12	13	14	15	16	17	18	19	20
21	22	23	24	25	26	27	28	29	30
31	32	33	34	35	36	37	38	39	40
41	42	43	44	45	46	47	48	49	50
51	52	53	54	55	56	57	58	59	60
61	62	63	64	65	66	67	68	69	70
71	72	73	74	75	76	77	78	79	80
81	82	83	84	85	86	87	88	89	90
91	92	93	94	95	96	97	98	99	100

BE CONSISTENT BE PERSISTENT

THE KEY TO ACHIEVING YOUR GOALS

I can use addition and subtraction to solve problems.
I can solve problems without giving up.

PYRAMID PROBLEMS

Can you fill in the empty blocks? Each block is the sum of the two blocks directly below it. You are to include positive whole numbers only, and not use a number more than once.

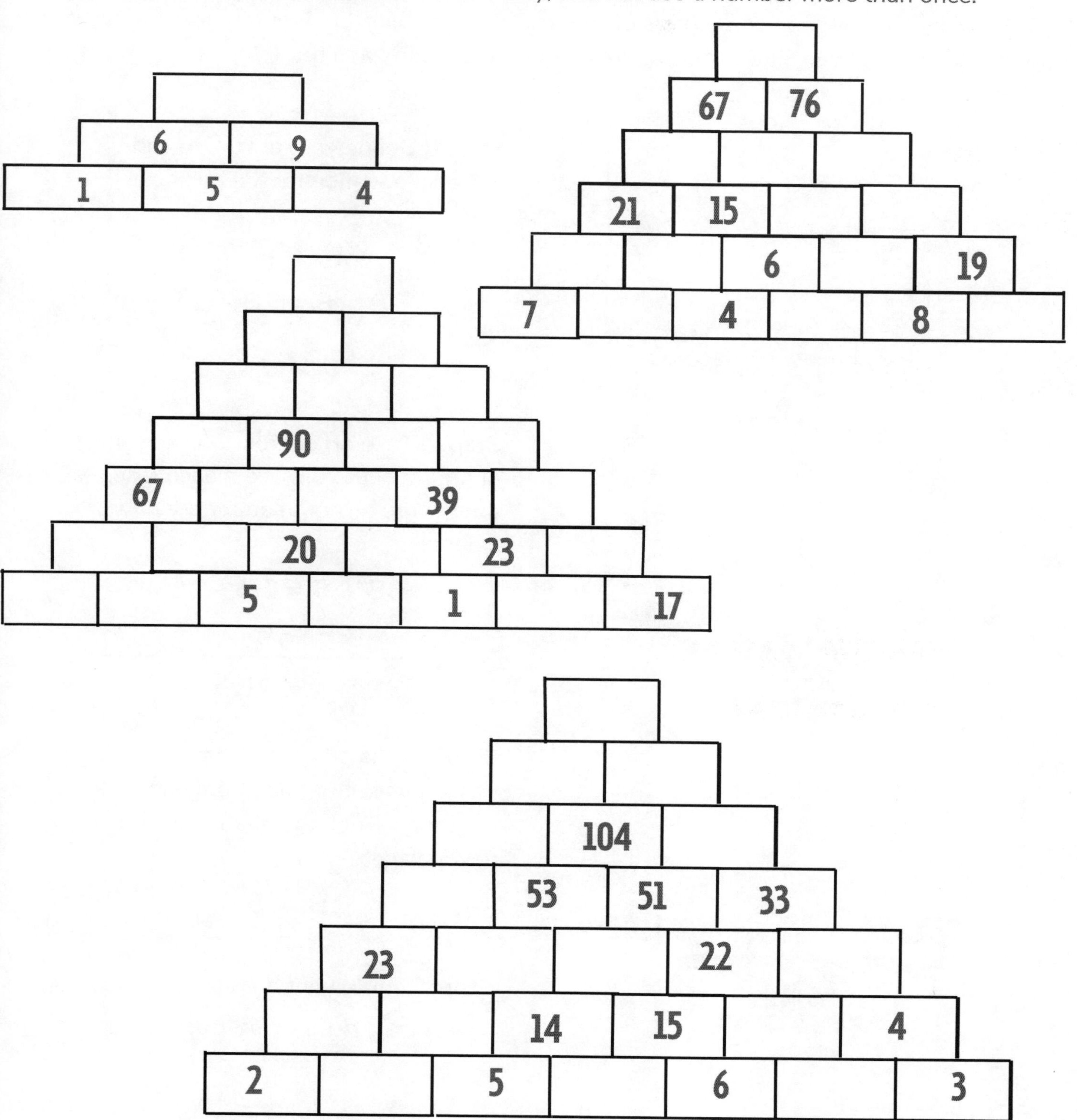

BALANCE PROBLEM example

The balance below has some bags of money and euro bills. Each bag holds the same number of euro bills. Find out how many euro bills are in each bag.

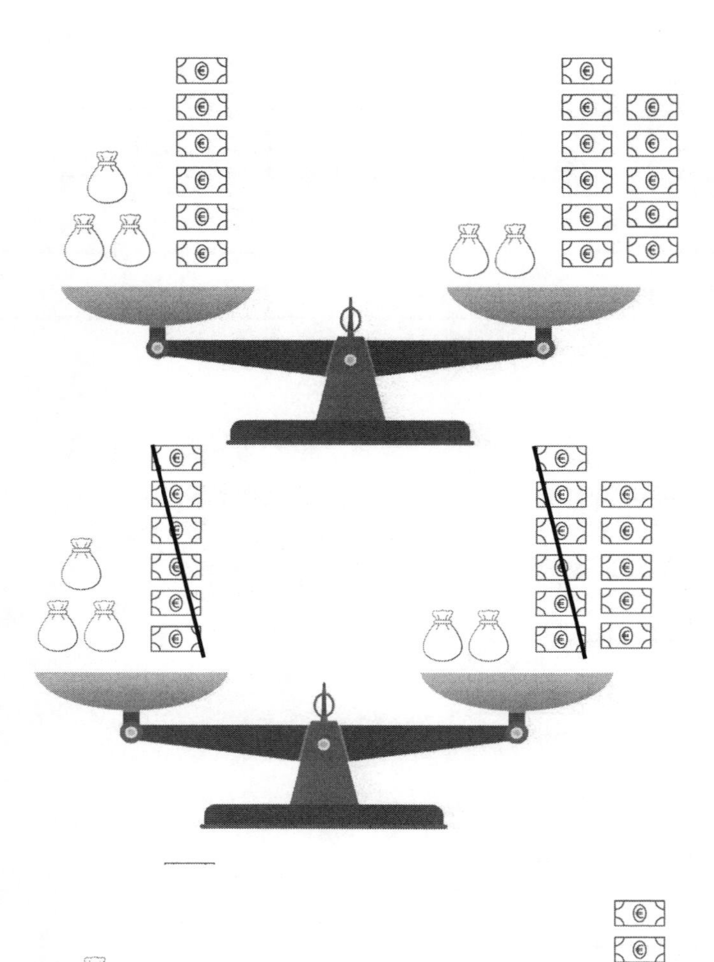

Step 1: Define your unknown and then write an equation for the balance.

Let B = one bag

$$3B + 6 = 2B + 11$$

Step 2: Take off or cross off the same number of euro bills from each side, then write an equation for the new balance.

$$\begin{array}{rcl} 3B + 6 & = & 2B + 11 \\ -6 & = & -6 \\ \hline 3B & = & 2B + 5 \end{array}$$

Step 3: Take off or cross off the same number of bags from each side, then write an equation for the new balance.

$$1B = 5$$

Step 4: Write your answer in a sentence.

Each bag holds five euro bills.

BALANCE PROBLEM YOUR TURN #1

The balance below has some bags of money and euro bills. Each bag holds the same number of euro bills. Find out how many euro bills are in each bag.

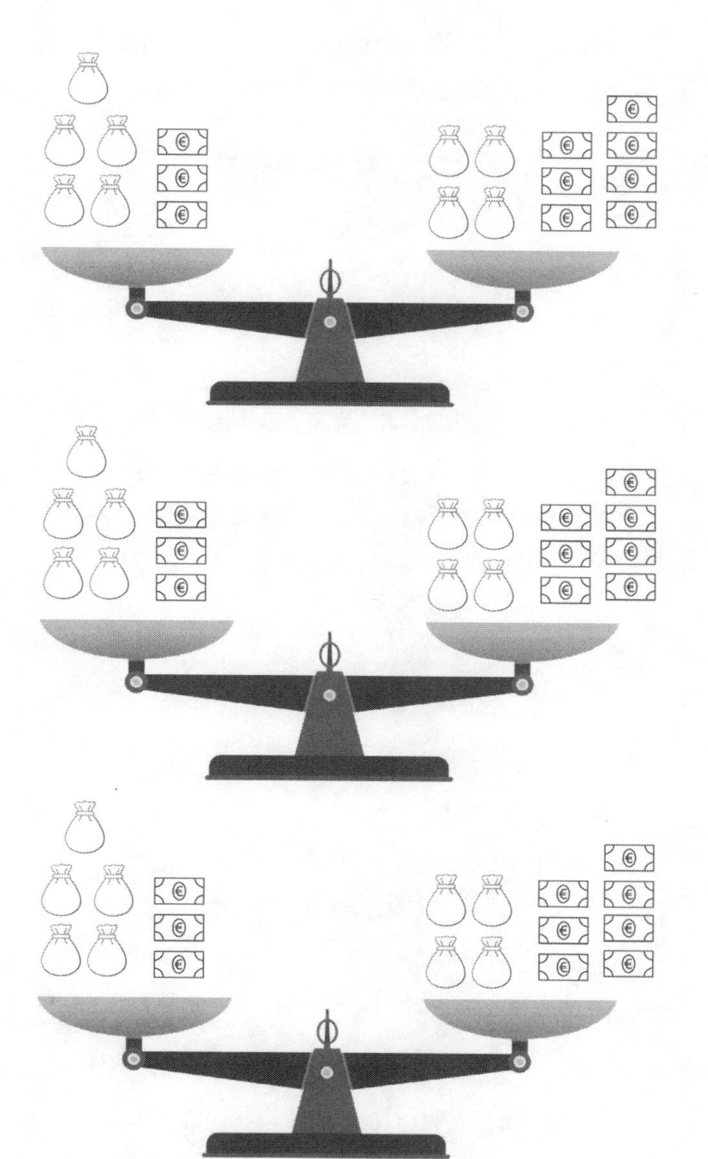

Step 1: Define your unknown and then write an equation for the balance.

Let B = one bag

Step 2: Take off or cross off the same number of euro bills from each side, then write an equation for the new balance.

Step 3: Take off or cross off the same number of bags from each side, then write an equation for the new balance.

Step 4: Write your answer in a sentence.

BALANCE PROBLEM YOUR TURN #2

The balance below has some bags of money and euro bills. Each bag holds the same number of euro bills. Find out how many euro bills are in each bag.

Step 1: Define your unknown and then write an equation for the balance.

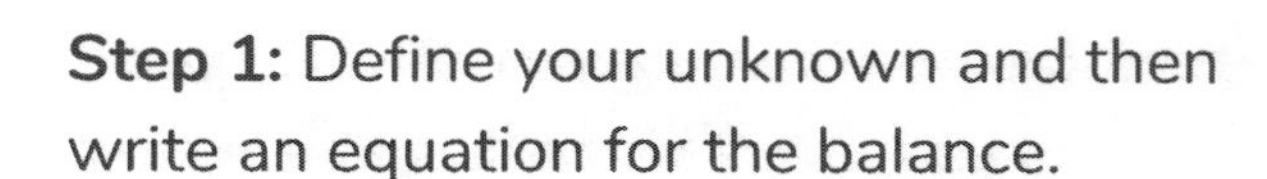

Let B = one bag

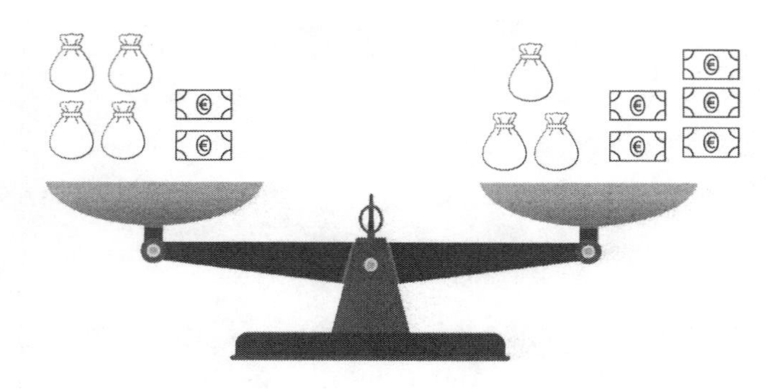

Step 2: Take off or cross off the same number of euro bills from each side, then write an equation for the new balance.

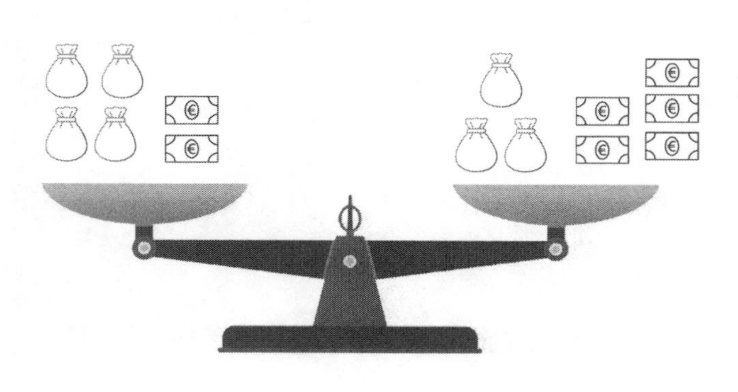

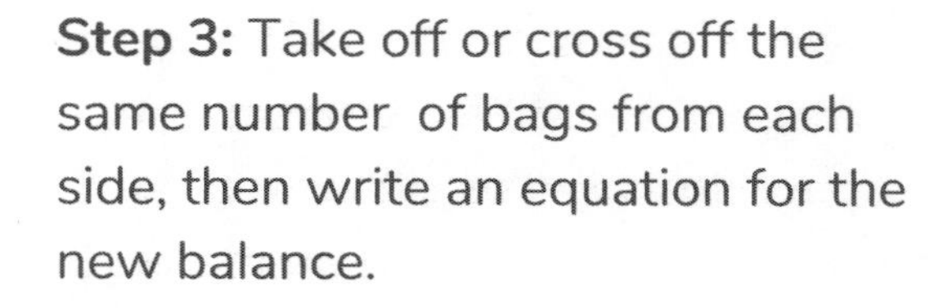

Step 3: Take off or cross off the same number of bags from each side, then write an equation for the new balance.

Step 4: Write your answer in a sentence.

Activity 61
1-4 CREATIVE THINKING

I can hold a growth mindset and solve open ended problems.
I can solve problems without giving up.

IMAGINATION WORKOUT #3

Imagine what the design could be, then finish the drawing yourself!
Give your drawing a title.

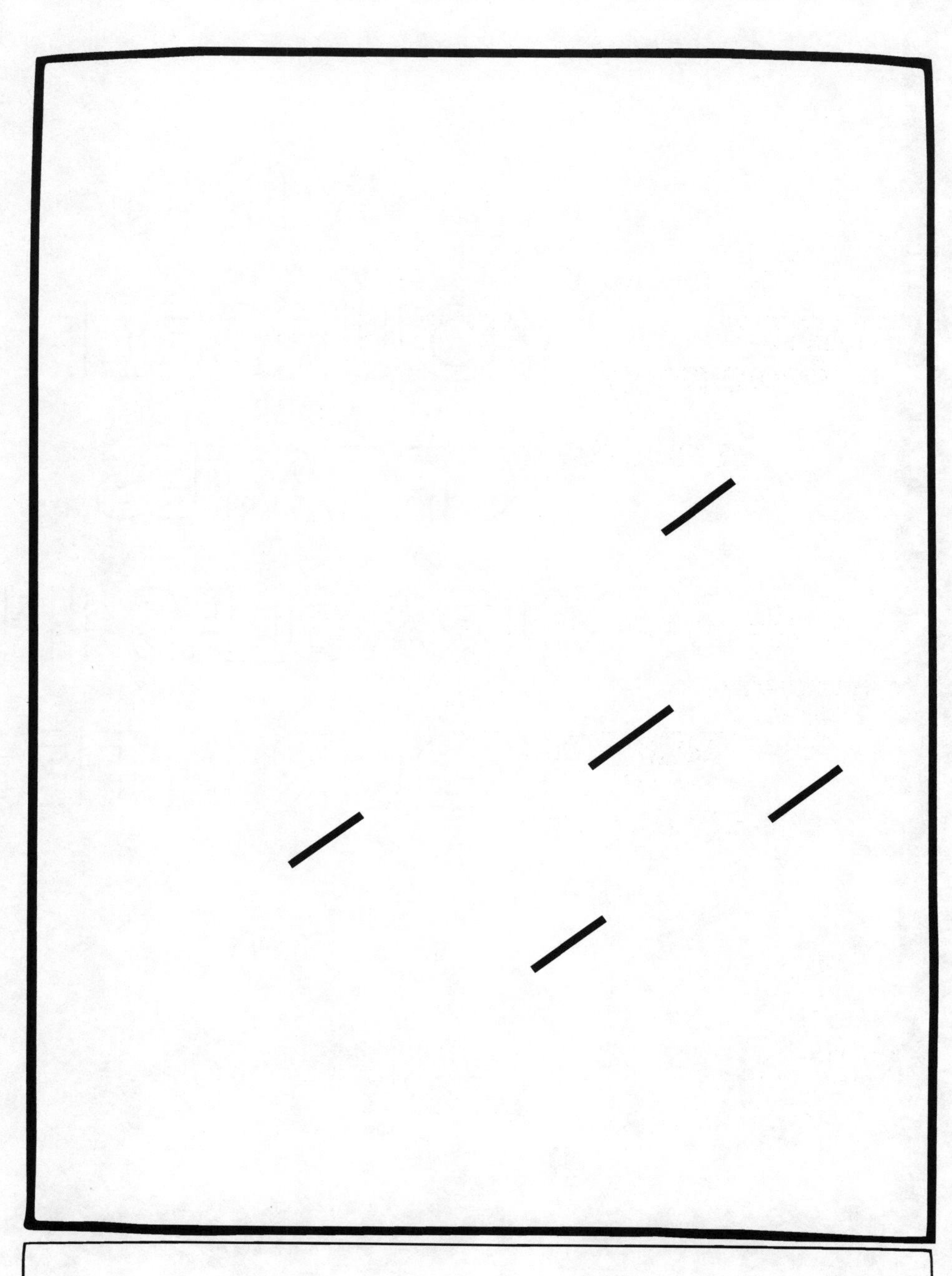

Activity 62
3-5 MINDFUL COLORING &
SOCIAL EMOTIONAL LEARNING

I can feel good about myself.
I can see the positive in my life.
I can focus on how to choose and apply color in a design to bring an awareness to the present moment.

Activity 63
3-5 GEOMETRY & PROBLEM-SOLVING

I can reason spatially

TANGRAM

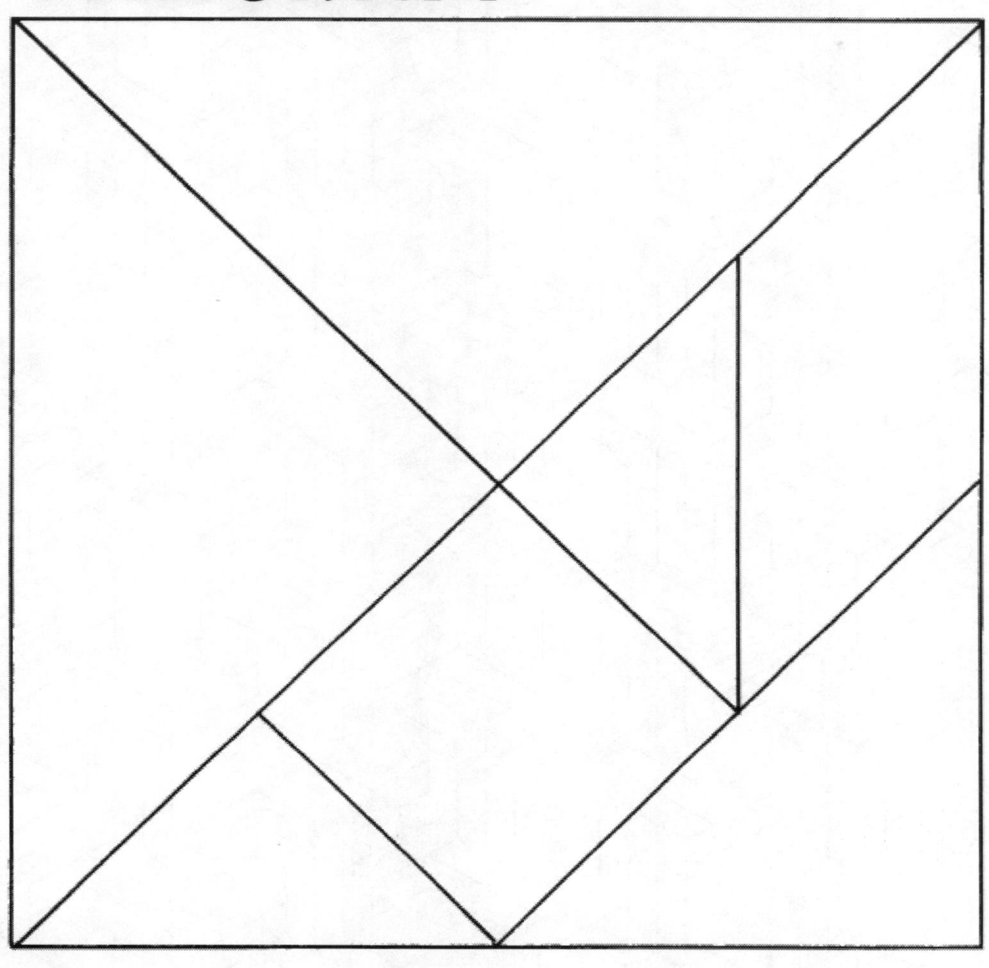

Cut out the seven shapes along the black lines, then arrange the seven pieces to make the three pictures below.

What other interesting pictures can you make with these seven pieces?

Activity 64
3-5 MINDFUL COLORING &
SOCIAL EMOTIONAL LEARNING

I can feel good about myself.
I can see the positive in my life.
I can focus on how to choose and apply color in a design to bring an awareness to the present moment.

Some things
have to end for
better things
to begin

Many thanks for purchasing this book. If you enjoyed this title then feedback on Amazon would be greatly appreciated. If you are not satisfied with this book, then please drop us an email at robin@acorns2oakes.com and we will do our best to sort the problem.

Scan this QR code to gain access to bonus content, and to check in with the solutions for this book.

Made in the USA
Monee, IL
20 October 2024

68343876R00059